Jeremy Cook

The Hatfield SCT Lunar Atlas

Photographic Atlas for Meade, Celestron and Other SCT Telescopes

With 216 Figures

 Springer

Jeremy Cook, BSc (Eng)

British Library Cataloguing in Publication Data
The Hatfield SCT lunar atlas:photographic atlas for
   Meade, Celestron and other SCT telescopes
   1. Moon–Maps
   I. Cook, Jeremy, 1933–2003
   523.3′0223
ISBN 1852337494

Library of Congress Cataloging-in-Publication Data
Cook, Jeremy, 1933–
   The Hatfield SCT lunar atlas:photographic atlas for Meade, Celestron and other SCT telescopes / Jeremy Cook.
      p. cm.
   Rev. ed. of: Amateur astronomer's photographic lunar atlas / Henry Hatfield. 1968.
   Includes bibliographical references.
   ISBN 1-85233-749-4 (hc:alk. paper)
  1. Moon—Maps. 2. Moon—Remote-sensing images. 3. Moon—Photographs from space. I. Hatfield, Henry.
Amateur astronomer's photographic lunar atlas. II. Title.

G1000.3.C6 2004
523.3′022′3—dc22

ISBN 1-85233-749-4 Springer-Verlag London Berlin Heidelberg
Springer Science+Business Media
springeronline.com

© British Astronomical Association 2005
Printed in China

Typeset by EXPO Holdings Sdn Bhd, Malaysia
58/3830-543210 Printed on acid-free paper SPIN 10930038

Jeremy Cook 1933–2003

The *Hatfield Lunar Atlas* has been widely used ever since it was first published. There has long been a need for a new edition, oriented to make it convenient for modern telescopes, and the obvious person to carry out this task was Jeremy Cook. Sadly, it was to be his last contribution to science.

He never worked as a professional astronomer, but he made his name as a skilled observer and became an acknowledged expert in all matters connected with the Moon. He was appointed Director of the Lunar Section of the British Astronomical Association and proved to be an exceptionally capable administrator, always ready to help others with the benefit of his experience. Jeremy was universally liked and admired, and his sudden and totally unexpected death came as an appalling shock to his many friends – and I am proud to have been counted as one of these. His marriage to Marie was long and ideally happy. She is herself a very capable lunar observer, and their son, Dr. Tony Cook, is an internationally known astronomer.

Jeremy Cook is much missed – but he is certainly not forgotten.

Sir Patrick Moore **CBE FRS**

Jeremy Cook (centre) with Commander Henry Hatfield, on the occasion
of the presentation of copies of *The Hatfield Photographic Lunar Atlas*,
outside Commander Hatfield's solar observatory.

Publisher's Preface

Jeremy was only a fortnight from delivering me the manuscript for this book when he died suddenly, just before Christmas 2003. He will be greatly missed. Jeremy was one of those people who always seemed to arrive in my office grin first. Unfailingly good-humoured, he was meticulous in preparing his work and was also one of that small band of authors who really cared about deadlines. We were both very enthusiastic about this project, and I'm devastated that Jeremy won't see the finished book.

The genesis of the idea was slightly unusual.

I am personally not very familiar with the Moon, and using *The Hatfield Photographic Lunar Atlas* to locate a particular crater proved next to impossible with my Meade LX-200 because of the mental gymnastics needed to visualise a map with South at the top and East on the left, as North at the top and East on the left – a mirror image. In the end I was reduced to going indoors, scanning the relevant page of the atlas, flipping it, printing it and running back out to the telescope.

At almost exactly the same time, in a field near Heidelberg in Germany, Hubertus Riedesel – one of Springer's international editorial directors and a keen amateur astronomer – was having exactly the same problem looking at the Moon with *his* LX-200. We discovered the coincidence the next morning.

Jeremy was the obvious person to approach to solve our problem, and so *The Hatfield SCT Lunar Atlas* was born.

My thanks to Tony Cook for rescuing the work from Jeremy's hard drive and to Mrs. Marie Cook for permission to use it.

John Watson
Hampshire, England
October 2004

Introduction

When the first version of this Atlas was first published, men were within a year of setting foot on the Moon, and the exciting buildup to this event stimulated a great upsurge in lunar observations by professionals and amateurs alike. The original author of this atlas, Henry Hatfield, put to good use the many photographs he took of the lunar surface at this time with his homebuilt 30 cm (12-inch) Newtonian telescope. He realised how useful it would be for an amateur observer to have by his side an easy-to-use Atlas of the Moon to assist in locating and identifying the many features which could be seen through a telescope of similar size—and so the *Amateur Astronomer's Photographic Lunar Atlas* came into being.

Thirty years later, the opportunity arose for the Atlas to be updated, not in the quality of the photographs (which remain excellent) but with the subsequent changes in lunar nomenclature, the change from Imperial to SI units of measurement and the change from Classical to IAU east-west directions. In recognition of the original author, the revised publication was named *The Hatfield Photographic Lunar Atlas*. It retained the original image orientation of lunar South at the top but with East now designated to be on the left and West on the right.

With the increasing use of commercially-made telescopes having a more compact light path than the Newtonian, a further edition of this lunar atlas has become desirable because most of these compact telescopes have their usual viewing position at right angles to the telescope's optical axis, which results in an image reversal. This change in direction of the optical path has the effect, for northern hemisphere observers, of providing a lunar image in the eyepiece that has North at the top and South at the bottom, but with Mare Crisium on the left and Aristarchus on the right—contrary to the accepted IAU convention.

Clearly, identification of lunar features at the telescope is made very much more difficult when the actual image is reversed left-right relative to an IAU-correct comparison map or photograph. This new Lunar Atlas, *The Hatfield SCT Lunar Atlas*, has therefore been constructed specifically for users of such telescopes in that the maps and photographs now appear in the same orientation as that seen through a right-angle eyepiece, that is North at the top, South at the bottom, East on the left and West on the right. This is ideal for almost all commercially made Schmidt-Cassegrain or Maksutov telescopes. It should be noted, however, that viewing a lunar image from the rear of such a telescope along its optical axis—that is, without a right-angle diagonal at the eyepiece—still produces a conventional IAU orientation.

Authority—Nomenclature

Feature names have, where appropriate, been altered to those listed in the NASA Reference Publication 1097 *NASA Catalogue of Lunar Nomenclature* by Leif E. Andersson and Ewen A. Whitaker (1982), and include more recent amendments particularly whereby some lettered craters have acquired proper names. However, a few features which are not contained in the NASA Catalogue have been allowed to retain their earlier names for historical reasons. These names are shown enclosed by brackets on the Maps in this book and identified in the Index of Named Formations by an asterisk.

The Maps—Overlaps—Scale—Map Grids

Since this Atlas is aimed primarily at visual observers with catadioptric telescopes (whose view of the lunar surface through the right-angle eyepiece will usually be one in which North is at the top, South at the bottom, East on the left and West on the right) all of the maps and main plates are printed in this orientation. We of course realise that the appearance of the maps and plates will be unfamiliar to other lunar observers because of this reversal!

The Atlas is divided into sixteen sections, each of which is made up of a map and two photographic plates. Each map is based primarily on the large facing plate. The Index of Named Formations includes many cross-references to help in locating the most appropriate map. Feature heights are given in metres (m) and distances in kilometres (km).

No attempt has been made to adhere rigidly to the boundaries of the various numbered sections in the Keys to Maps and Plates: indeed in some cases the boundaries of the plates have purposely been allowed to encroach into neighbouring sections. The Key Plate is intended to guide the reader into the right area: the Index of Named Formations lists every map on which a particular feature will be found. The "scale" of each map, and of the main plates which accompany it, has been adjusted so that the Moon's diameter is nominally 64 centimetres (any variation in this diameter is stated in the relevant caption). Despite this, the true scale (the relationship between the diameter of a crater on the Moon and its diameter in this Atlas) of each map and plate varies from place to place; furthermore the north-south scale may well vary in a different way from the east-west scale. This is part of the very nature of the orthographic projection, in which the observer views a globe from a great distance. In order to give the reader some idea of the true scale from place to place, the diameters of five craters have been noted beneath each map. As the scale varies from plate to plate, so the size and shape of the various photographic images will also vary. The grids on the various maps are intended to be used only for reference purposes. They bear no relationship to lunar latitude and longitude or to any of the cartographic grids which are in use at the present time. If a formation appears on more than one map, then its grid references on each map will almost certainly vary. An unnamed and unlettered formation on any map may be identified with certainty by quoting the map number and grid reference, e.g., Map 3 Square a4, and then making a small tracing of the square concerned, showing the formation.

The Libration Keys

Beneath each main plate there is a small key, which shows the numbered area in which the plate lies and the Moon's optical libration when the photograph was taken. The Moon does not present exactly the same face to the Earth all the time, but rocks gently back and forth in all directions, so that at one time or another an observer on the Earth will see about 60% of its surface. Referring to these libration keys, the reader should imagine that the Moon has rotated in the direction of the arrow by the amount indicated. Thus in Plate la the movement is 7.3° in a direction a little east of north, and therefore an area just west of the South Pole has been exposed, which could not be seen if the Moon were in its mean position. If the blacked-in "square" on these keys lies in the same semicircle as the libration arrow, then the direction of the libration is fairly favourable for the area depicted on the plate; if the arrow passes through the square concerned, then the direction is very favourable; an unfavourable libration will exist if the square concerned lies beyond the head of the libration arrow. The amount of libration can vary between nothing and a maximum of about 10°. Libration of 7° or 8° is considered to be good in most cases. The Moon's physical libration and diurnal libration have been ignored since their combined effect would not alter the aspect of the various photographs appreciably.

Conventional IAU Plates

The small plates show the same section of the Moon as the corresponding main plate, but with the IAU-correct orientation: that is, South at the top, North at the bottom, East on the left and West on the right. Once a feature has been located in the eyepiece and identified using the large plates and maps, it is relatively easy to find it on the small plates. This is intended to help users of modern commercial telescopes become familiar with the location of features on convential Moon maps.

Further Information

The photographs are typical of those taken through a Newtonian telescope of 300 mm (12-inch) aperture. In the years since the Atlas was first published, spectacular advances have been made in lunar surface imaging by manned and unmanned spacecraft using photographic processes and CCD imaging in many optical wavebands. It is possible that water ice (or at least water molecules) may have been detected by the Clementine and Lunar Prospector spacecraft in the regions around the north and south poles, a hypothesis that would not have been considered seriously until comparatively recently.

The sophisticated equipment and platforms required to make such discoveries are currently well beyond the reach of amateur observers, but there is little doubt that some of these will become available in due course. However, in case anyone imagines that there is nothing new to discover on the lunar surface, do not forget that professional astronomers cannot observe the Moon in real time like an amateur observer, and spacecraft have a relatively short lifespan. Subtle changes have been observed to occur on the lunar surface, and prolonged study of these and other topographic features is an excellent way of learning the geology (or rather selenography) of our closest cosmic neighbour.

Users of this Atlas are strongly advised to join a local astronomical society or club, where they will obtain advice and assistance in pursuing their interest.

I would like to express my grateful thanks to the Council of the BAA (who own the original copyright) for allowing the use of the maps and photographs which form the basis of the present publication. I am also grateful for advice and assistance freely given by Ewen Whitaker and by the original author, Henry Hatfield.

British Astronomical Association   Jeremy Cook
Burlington House   December 2003
Piccadilly, London, England

Maps and Plates

Keys to Maps and Plates

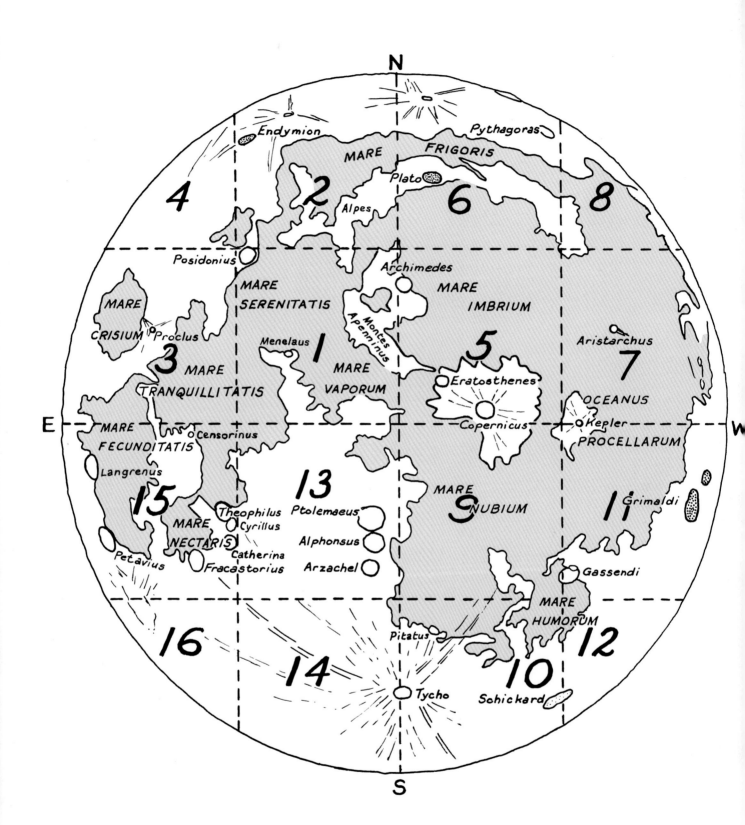

0240 UT 05.08.63

Map 1

Crater Diameters

Archimedes	83 km	(h7)
Posidonius	95 km	(b7)
Manilius	39 km	(e4)
Maskelyne	24 km	(a1)
Bruce	7 km	(g1)

This map has been prepared from Plate 1a. Plates 1b, 1c, and 1d show the same ar under different lighting. Plate 1e shows larger scale photographs of the Triesneck (g2) and Linné (e7) areas.

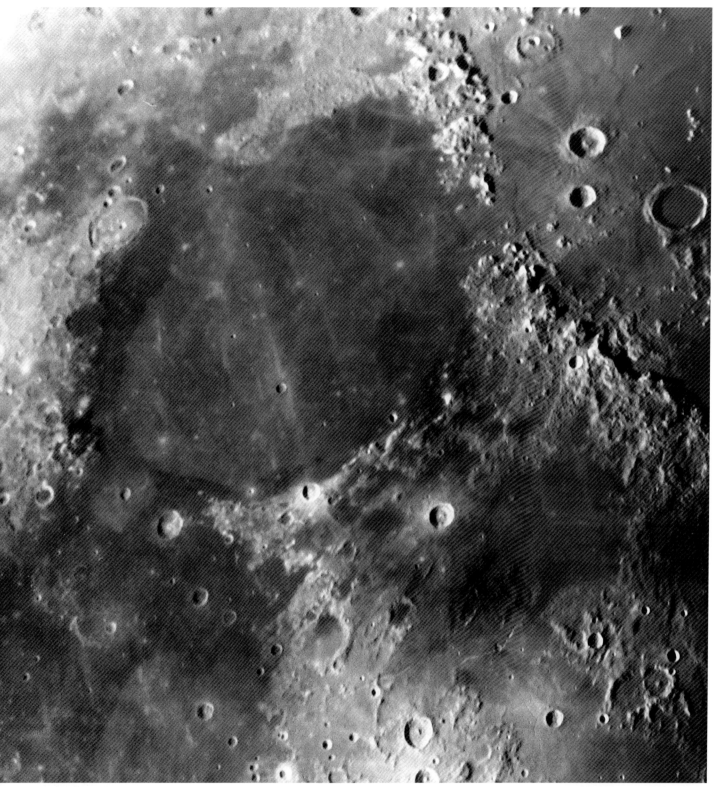

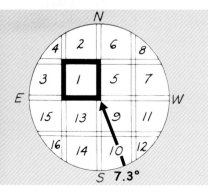

2046 UT 28.04.66. Moon's age 7.9 days. Diameter 64 cm. The Montes Appeninus (g5) rise to 4900 m in places.

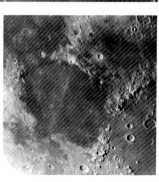

Plate 1b

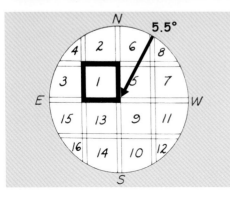

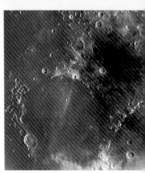

0215 UT 06.08.66. Moon's age 18.9 days. Diameter 64 cm. Note the ridges on the mare to the south west of Posidonius (b7), and the domes near Arago (c2). These are shown on a larger scale on Plate 3f.

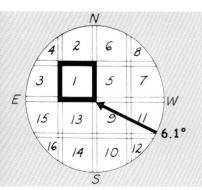

2137 UT 25.12.66. Moon's age 13.8 days (2 days before Full Moon). Diameter 64 cm. Compare this with Plate 1a, which shows almost the same area.

Plate 1d

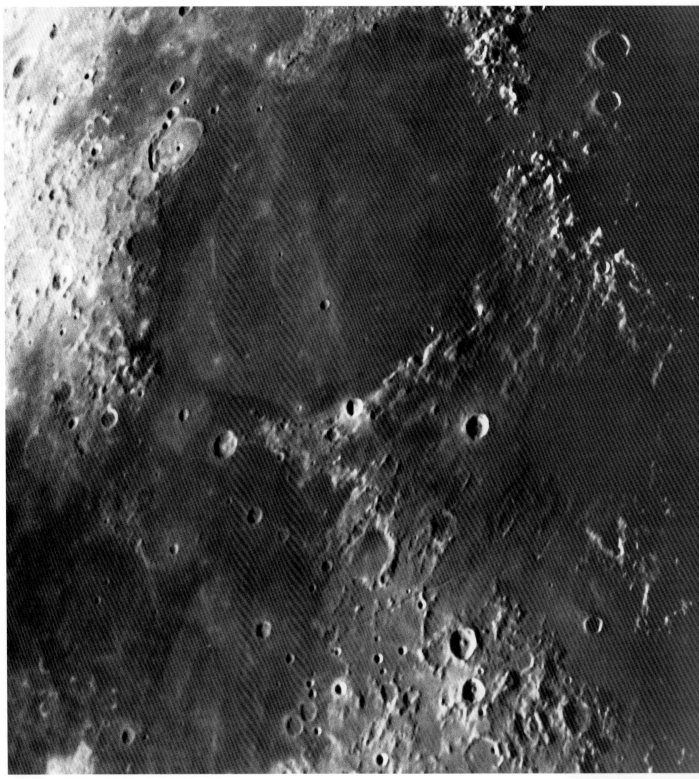

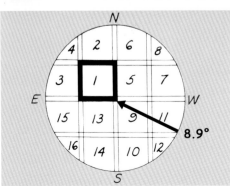

1835 UT 18.03.67. Moon's age 7.6 days. Diameter 64 cm. Although the Moon's age here is almost the same as in Plate 1a, the aspect is different, because of the different libration.

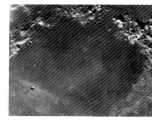

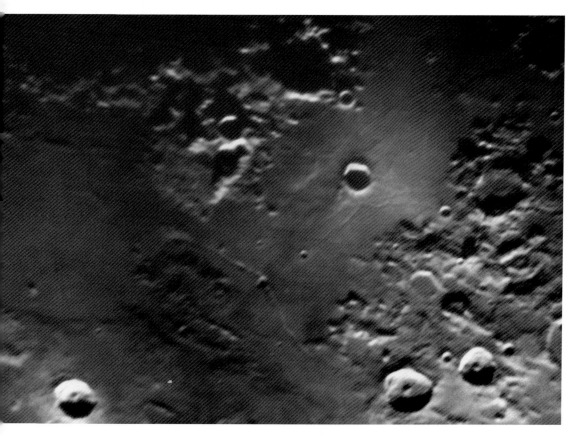

Linné, Bessel and Sulpicius Gallus in the Mare Serenitatis. 2017 UT 16.05.67. Moon's age 7.2 days. Diameter 94 cm approx. Linné lies in Map 1 e7.

Triesnecker, Hyginus and their systems of rilles. 2016 UT 16.05.67. Moon's age 7.2 days. Diameter 94 cm approx. Triesnecker lies in Map 1 g2.

Map 2

Crater Diameters

Autolycus.....................................39 km (f1)
Daniell ....29 km (a2)
Egede ....37 km (e4)
Strabo ....55 km (b7)
Gioja ....42 km (f8)

This map has been prepared from Plate 2a. Plates 2b, 2c and 2d show the same ar
under different lighting. Plate 2e shows larger scale photographs of the Vallis Alp
(f4) area.

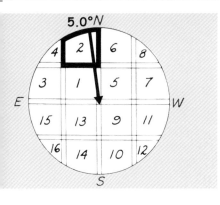

1814 UT 22.11.66. Moon's age 10.2 days. Diameter 64 cm. This is a very favourable northerly libration, which is not likely to be seen very often. Note that Plato (h5) is very nearly circular. The Montes Alpes are about 3700 m high in places.

Plate 2b

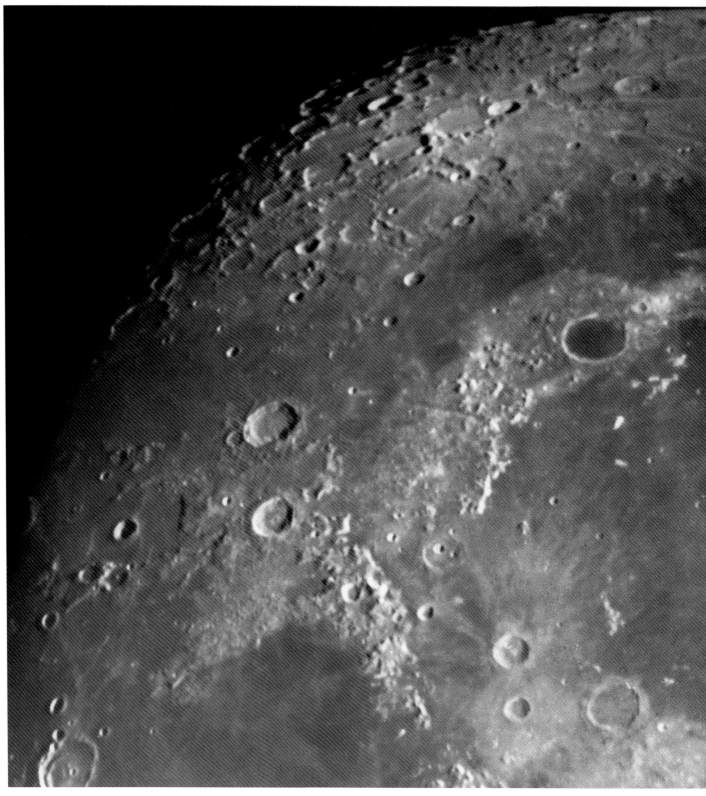

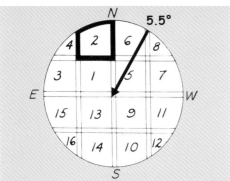

0211 UT 06.08.66. Moon's age 18.9 days. Diameter 64 cm. Compare this libration with that of Plate 2c.

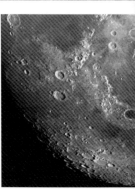

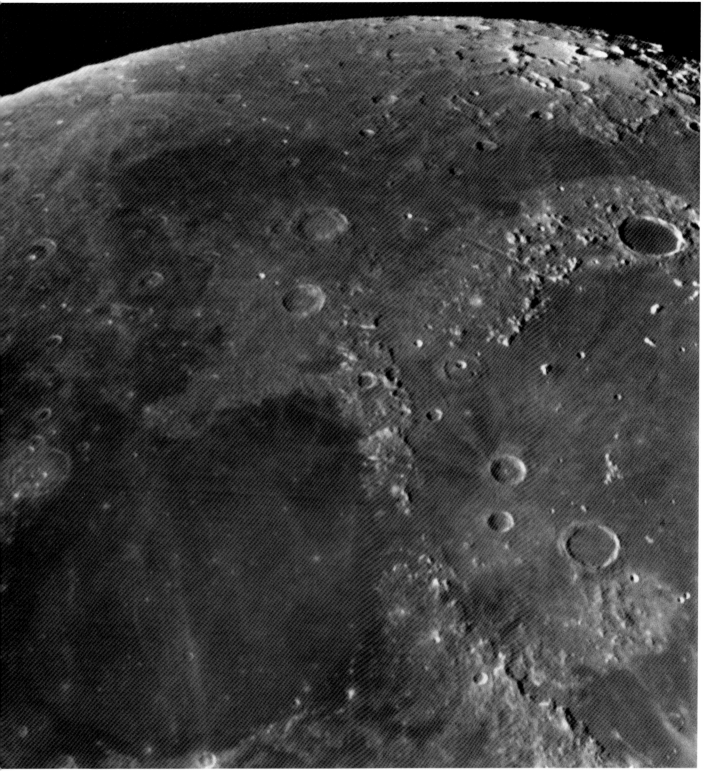

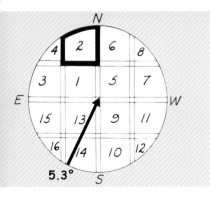

2103 UT 29.05.66. Moon's age 9.4 days. Diameter 64 cm. This is a relatively bad libration for this area. Note how much closer Plato (h5) is to the limb here than in Plates 2a and 2b.

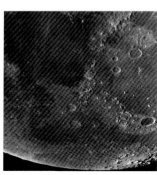

Plate 2d

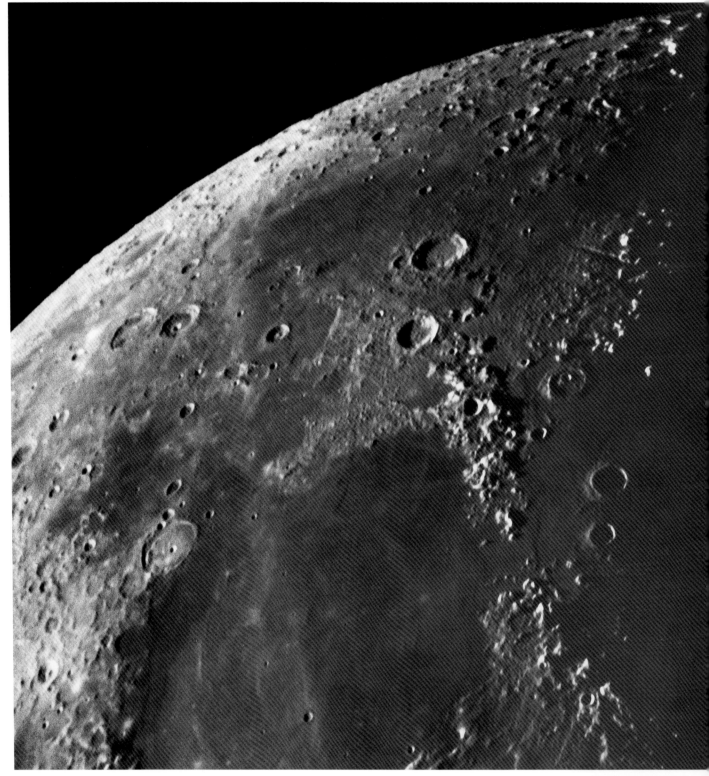

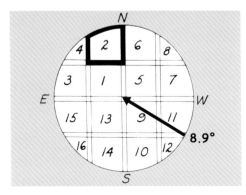

1835 UT 18.03.67. Moon's age 7.6 days. Diameter 64 cm. Note how Mons Piton (g3) catches the early morning sunlight. The Montes Caucasus (e2) rise to about 6000 m.

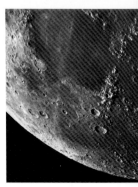

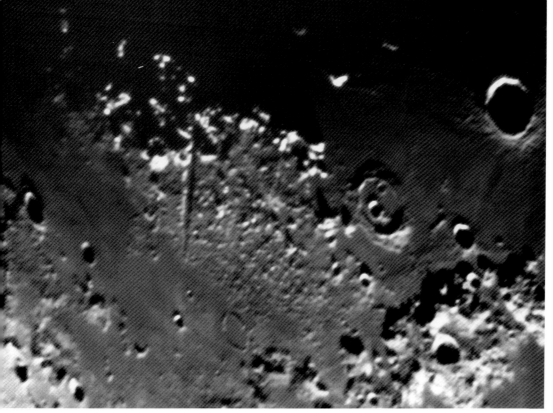

Cassini, the Vallis Alpinus and Plato. 1925 UT 19.03.67. Moon's age 8.6 days. Diameter 91 cm approx. Note the "ghost" crater ring just South of Plato. This has been called "Ancient Newton".

Aristillus, Cassini and the Vallis Alpinus. 2012 UT 16.05.67. Moon's age 7.2 days. Diameter 94 cm approx. The Vallis Alpinus lies in Map 2 f4.

Map 3

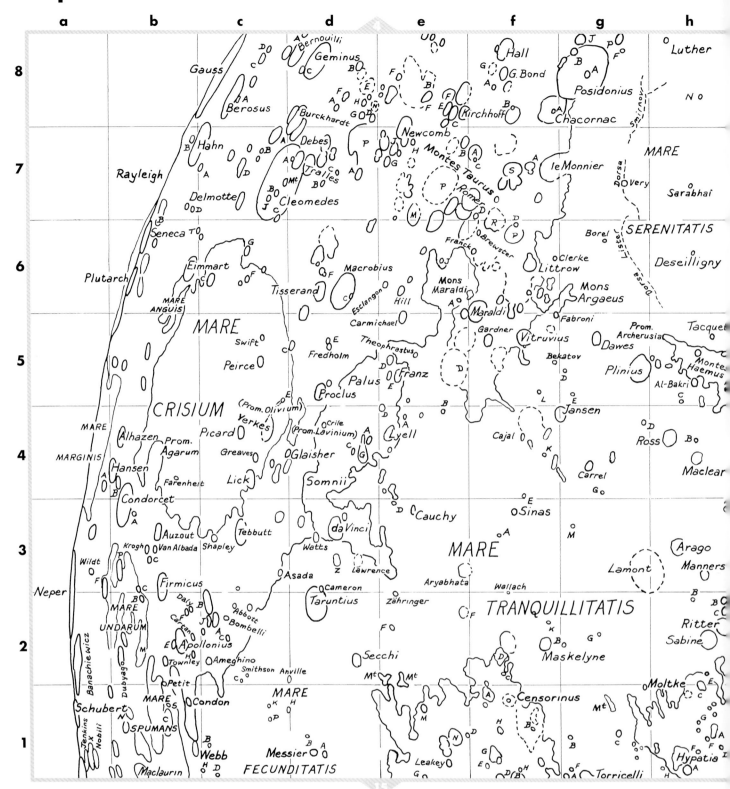

Crater Diameters

Luther .... 10 km (h8)
Berosus .... 74 km (c8)
Lyell.................................. 32 km (e4)
Sabine .... 30 km (h2)
Maclaurin .... 50 km (b1)

This map has been prepared from Plates 3a and 3d. Plates 3b, 3c and 3e show
same area under different lighting. Plate 3f shows larger scale photographs of
area near Arago (h3), and the Wrinkle Ridges east of le Monnier (g7). Plate 3g sh
Messier and Messier A (d1).

Note: In Plate 3c d4 are shown the two Promentaria Olivium and Lavinium wh
were once thought to be spanned by a natural arch. This is now known to u
shadow effect and the names are not in current use.

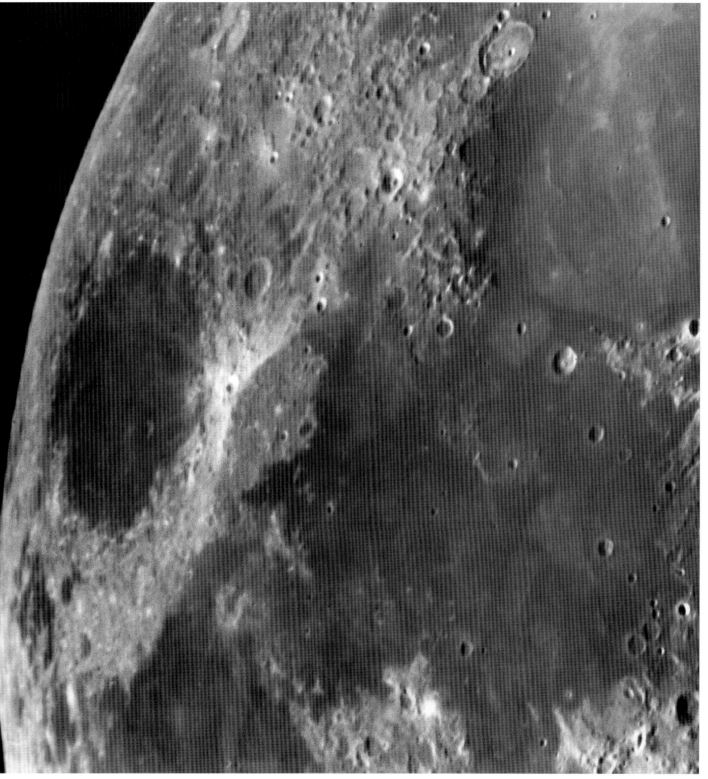

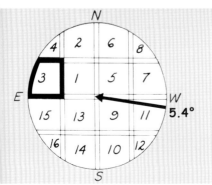

1757 UT 16.02.67. Moon's age 7.3 days. Diameter
64 cm. Note the rays round Proclus (d5). There are
several "domes" North and West of Arago (h3).

Plate 3b

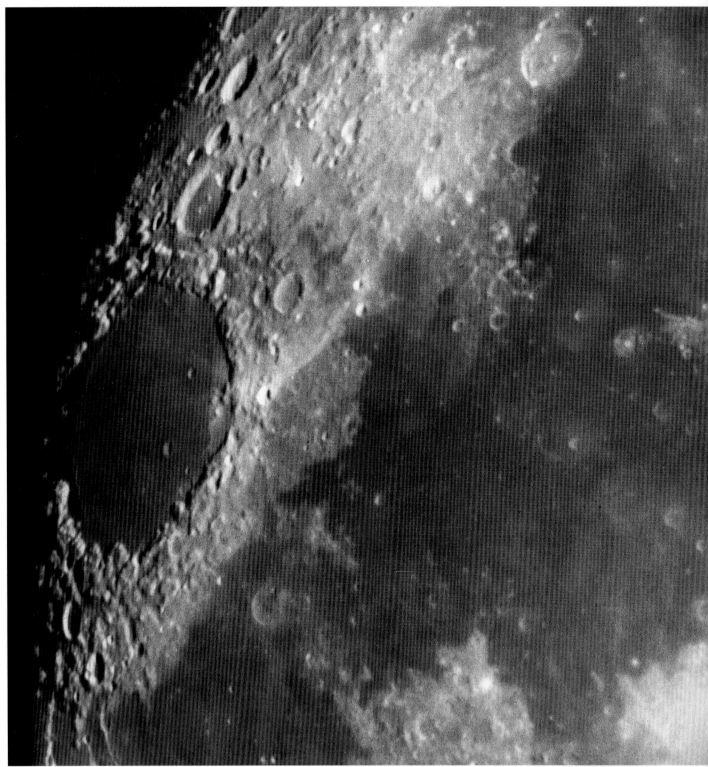

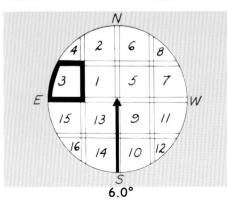

6.0°

2228 UT 08.01.66. Moon's age 17.0 days. Diameter 64 cm. Compare the shape of the Mare Crisium (c5) here with that in Plate 3a and the left hand photograph in Plate 3e.

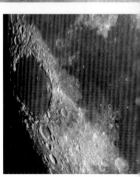

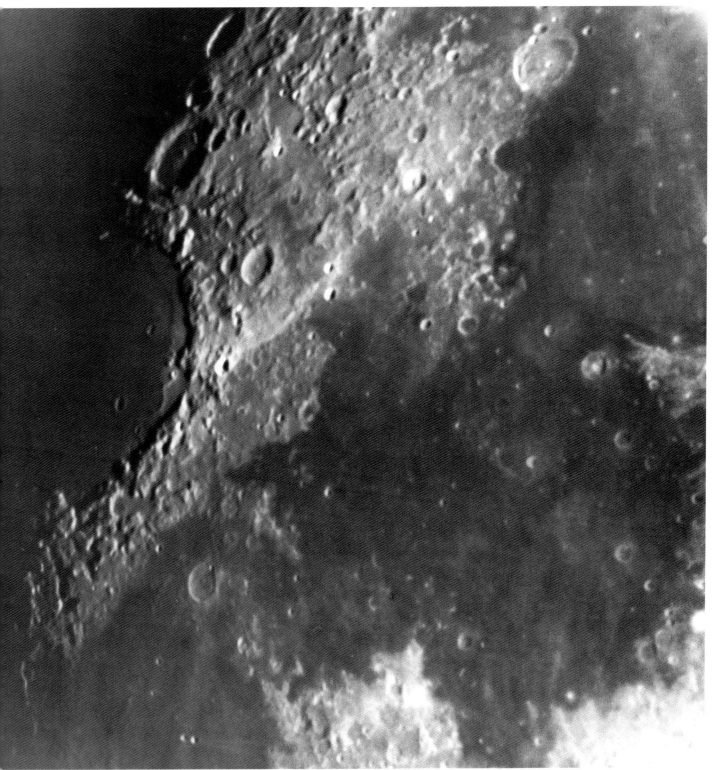

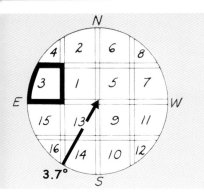

2336 UT 26.02.67. Moon's age 17.6 days. Diameter 64 cm. Compare the Mare Tranquillitatis (f3) here with the same area in Plate 3a.

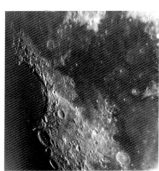

Plate 3d

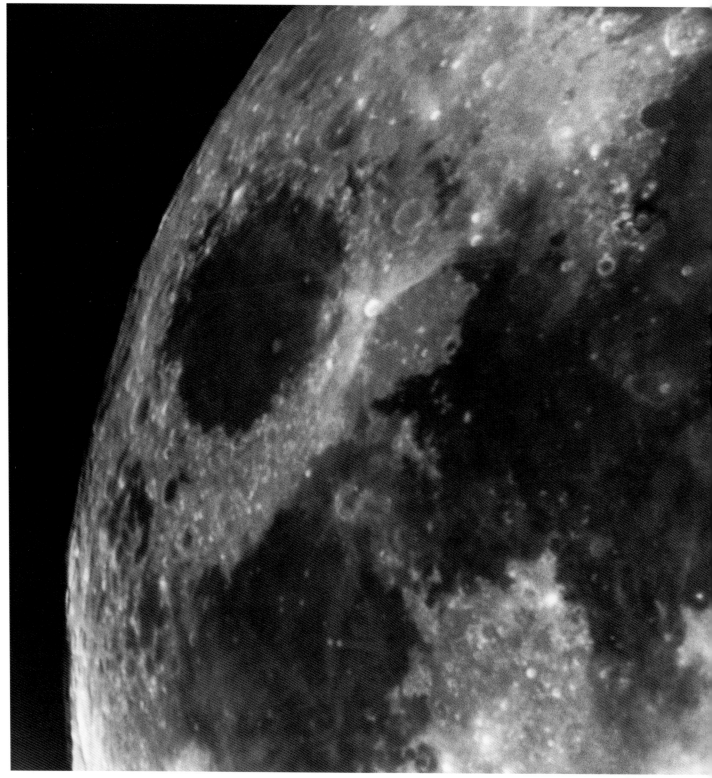

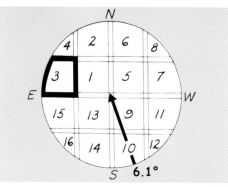

2353 UT 24.02.67. Moon's age 15.6 days. Diameter 64 cm. This was taken 6 hours after Full Moon.

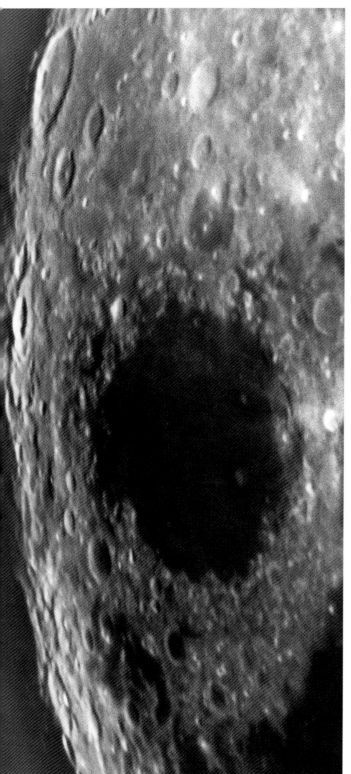

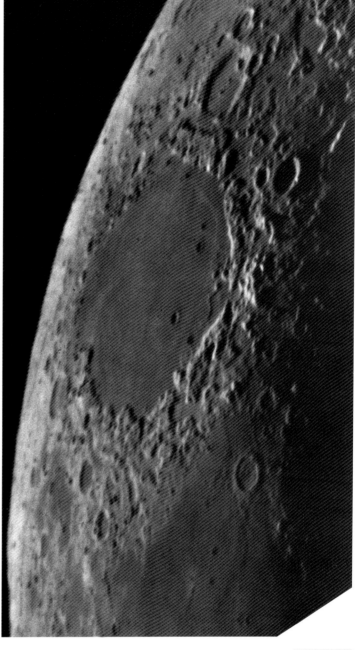

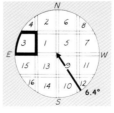

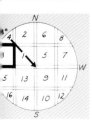

2304 UT 20.08.67.
Moon's age 14.9 days.
Diameter 64 cm.
Gauss (c8) is near the
limb and Neper (a2)
is on the limb.

2034 UT 23.05.66.
Moon's age 3.4 days.
Diameter 64 cm. Com-
pare this with Plate 3b
and with its neighbour
here.

Plate 3f

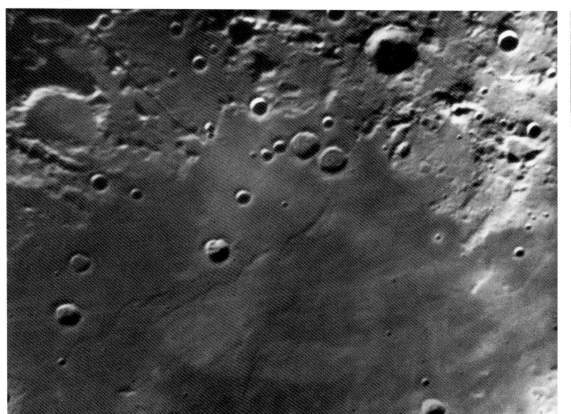

Arago, its "Domes" and Rima Ariadaeus. 2000 UT 15.05.67. Moon's age 6.2 days. Diameter 94 cm approx. Arago lies in Map 3 h3 and Map 1 c2. The domes lie about its own diameter North and West of it. Rima Ariadaeus lies in Map 1 d2.

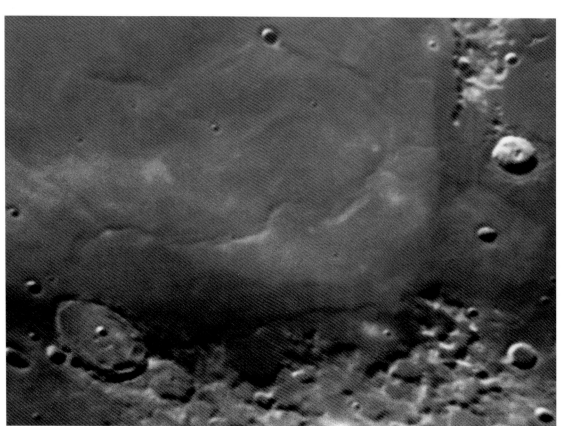

Plinius, Posidonius, Bessel and the "Wrinkle Ridges" in the Mare Serenitatis. 2015 UT 15.05.67. Moon's age 6.2 days. Diameter 94 cm approx. Plinius lies in Map 3 g5.

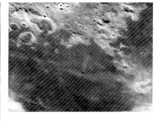

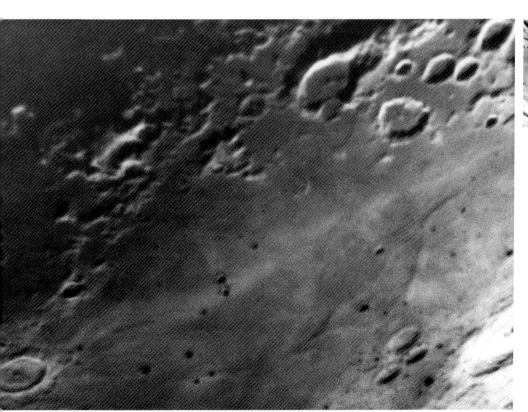

Messier and Messier A. 2008 UT 15.05.67.
Moon's age 6.2 days. Diameter 94 cm approx.

Note how these two craters change their appearance in two days. They lie in Map 3 d1.

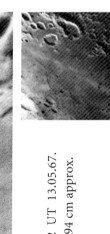

Messier and Messier A. 2002 UT 13.05.67.
Moon's age 4.2 days. Diameter 94 cm approx.

Map 4

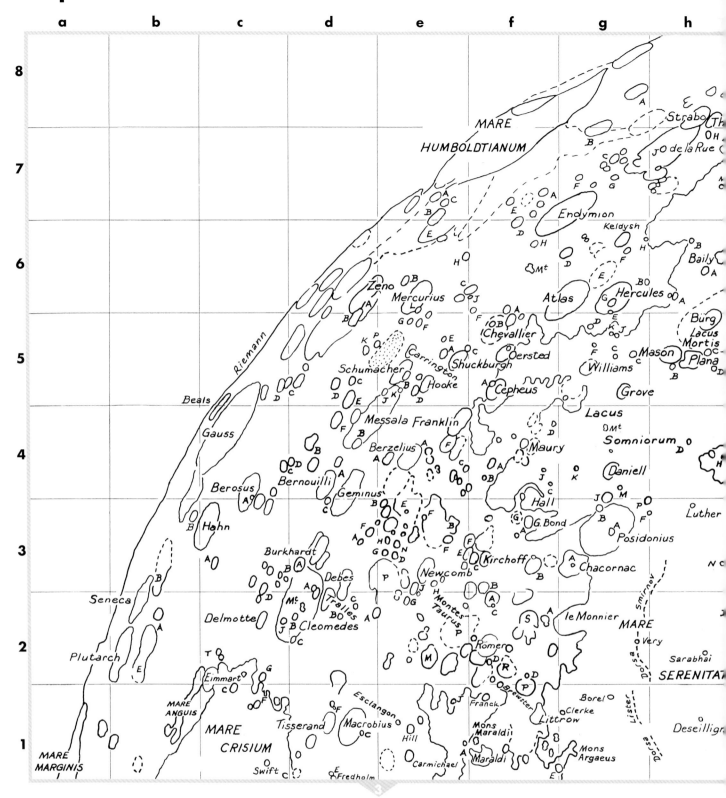

Crater Diameters

Endymion	125 km	(g7)
Mason	34 km	(h5)
Gauss	177 km	(c4)
Römer	40 km	(f2)
Tisserand	37 km	(d1)

This map has been prepared from Plates 4a and 4e. Plates 4b, 4c and 4d show the s
area under different lighting.

Note: Plate 4e shows quite a good libration for this area. The Mare Humboldtia
(f7) and Gauss (c4) do not often appear like this.

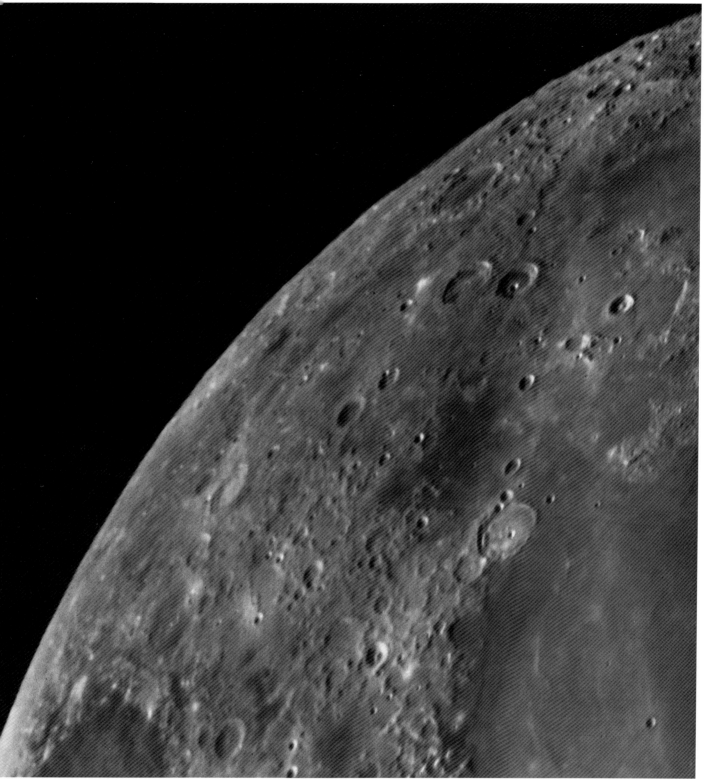

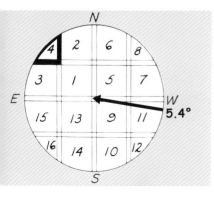

1757 UT 16.02.67. Moon's age 7.3 days. Diameter 64 cm.

Plate 4b

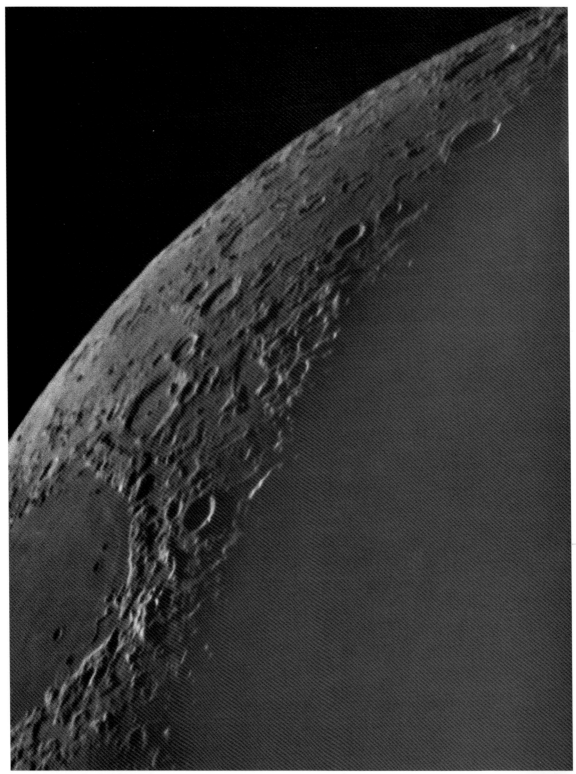

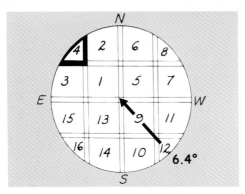

2034 UT 23.05.66. Moon's age 3.4 days. Diameter 64 cm. This is not a favourable libration for this area, Endymion (g7) lies almost on the limb.

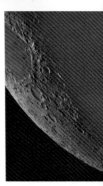

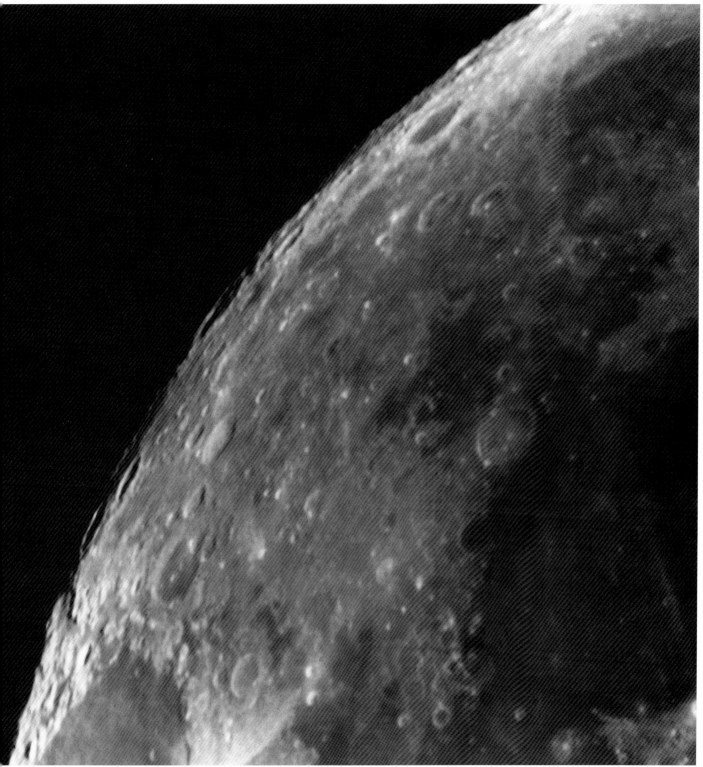

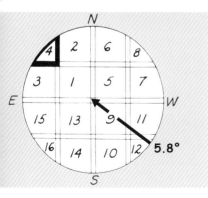

2127 UT 28.11.66. Moon's age 16.3 days. Diameter
64 cm. Libration is similar to that in Plate 4b.

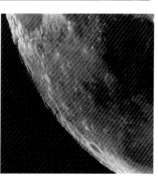

Plate 4d

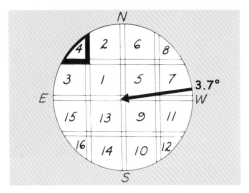

2248 UT 29.10.66. Moon's age 15.7 days. Diameter 64 cm. Compare this with Plate 4e. The area is the same in each case but the libration is very different.

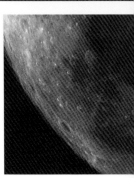

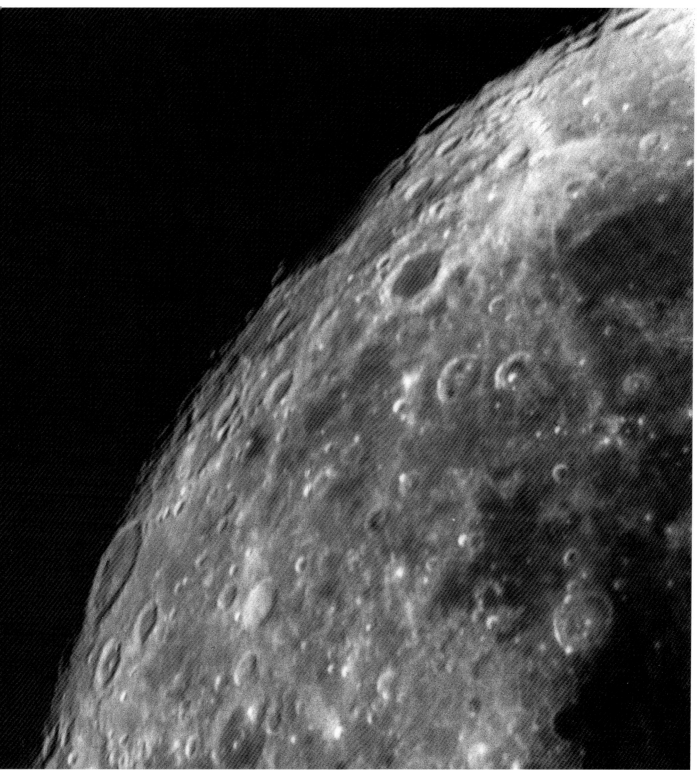

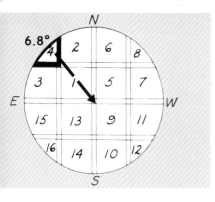

2304 UT 20.08.67. Moon's age 14.9 days. Diameter 64 cm. This is quite a good libration for this area. The Mare Humboldtianum (f7) and Gauss (c4) do not often appear like this.

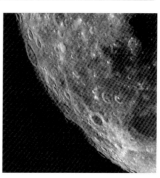

Map 5

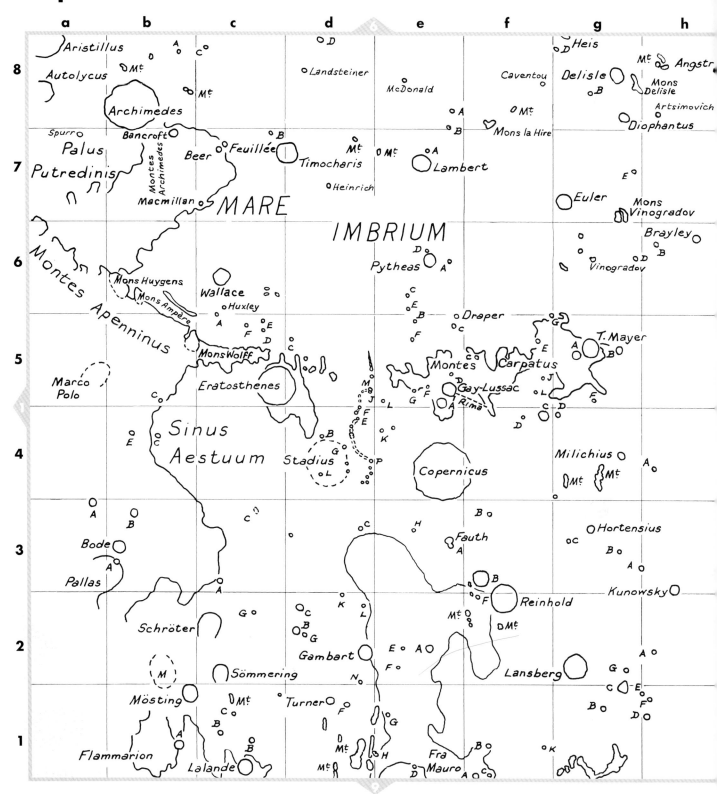

a   b   c   d   e   f   g   h

MARE

IMBRIUM

Montes Apenninus

Sinus Aestuum

Copernicus

Crater Diameters

Delisle	25 km	(g8)
Beer	10 km	(c7)
Copernicus	93 km	(e4)
Lansberg	39 km	(g2)
Mösting	25 km	(b1)

This map has been prepared from Plate 5a. Plates 5b, 5c and 5d show the same
under different lighting. Plate 5e shows larger scale photographs of the area ar
Copernicus (e4) and of the "Domes" in the vicinity of Milichius (g4).

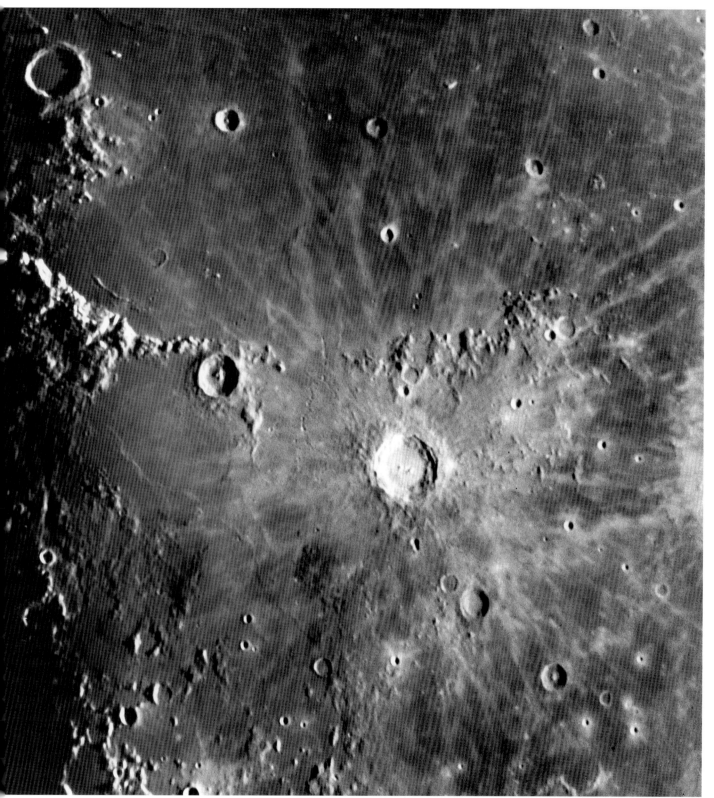

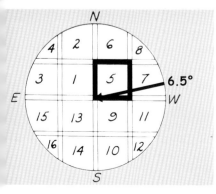

0315 UT 09.08.66. Moon's age 23.8 days. Diameter 64 cm. Note the ridges in the Mare Imbrium and the Sinus Aestuum (c4). The Montes Carpatus rise to about 2100 m in places.

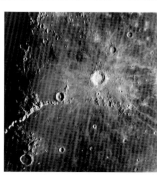

Plate 5b

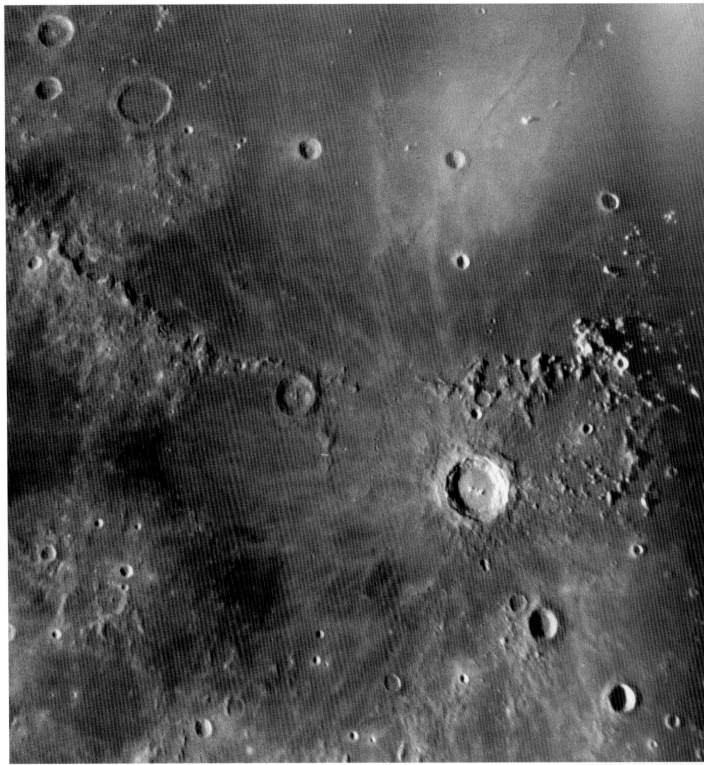

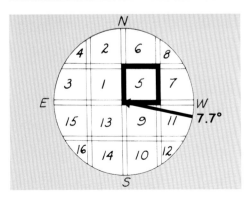

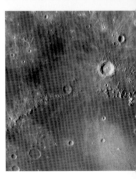

1947 UT 31.01.66. Moon's age 10.1 days. Diameter 64 cm. Compare this with Plate 5c which was taken nearer to Full Moon.

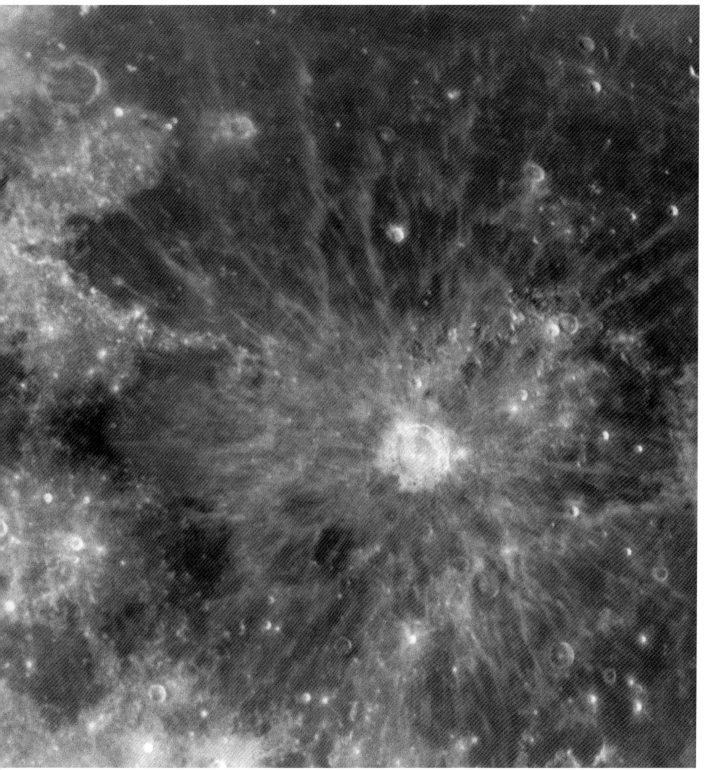

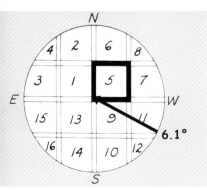

2135 UT 25.12.66. Moon's age 13.8 days. Diameter 64 cm. This photograph shows almost exactly the same area as Plate 5b.

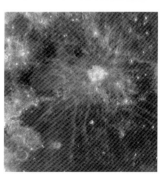

Plate 5d

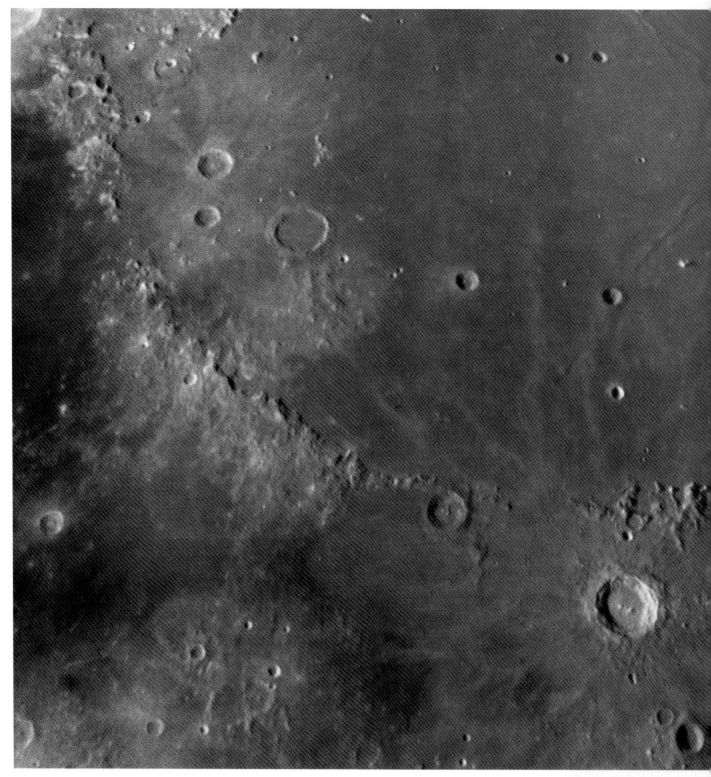

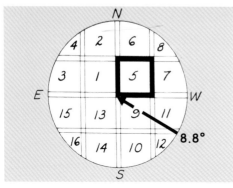

1801 UT 19.02.67. Moon's age 10.3 days. Diameter 64 cm. This photograph extends further to the East than the others in this group.

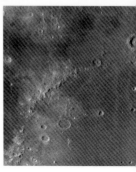

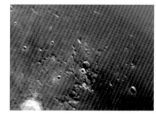

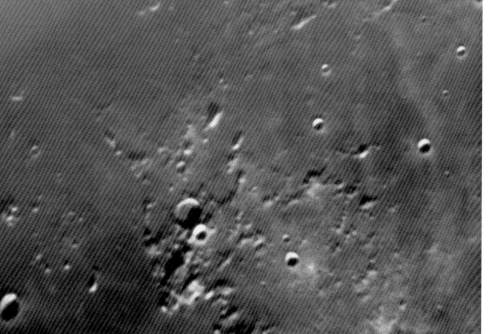

Euler to Hortensius, with the "Domes" near Milichius. 1908 UT 21.03.67. Moon's age 10.6 days. Diameter 91 cm approx. Milichius lies in Map **5** g4; there are several domes near it here, mostly to the North West.

The Copernicus area soon after sunrise. 2041 UT 20.03.67. Moon's age 9.7 days. Diameter 91 cm approx. Draper (e5, e6) is 8 km in diameter.

Map 6

Crater Diameters

Cleostratus	63 km	(h7)
Epigenes	55 km	(b7)
Le Verrier	20 km	(d4)
Angström	10 km	(h2)
Timocharis	34 km	(c1)

This map has been prepared from Plate 6a. Plates 6b, 6c, and 6d show the same under different lighting. Plate 6e shows larger scale photographs of the Montes (d5) and surrounding country, and of the Sinus Iridum (e4) soon after Sunrise.

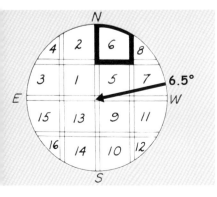

0315 UT 09.08.66. Moon's age 23.8 days. Diameter 64 cm. Note the "ghost" craters and low ridges near the Eastern border of the Mare Imbrium. Mons Pico (b4) is about 2400 m high and is not nearly so steep as it looks.

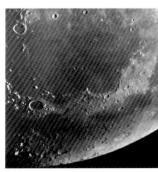

Plate 6b

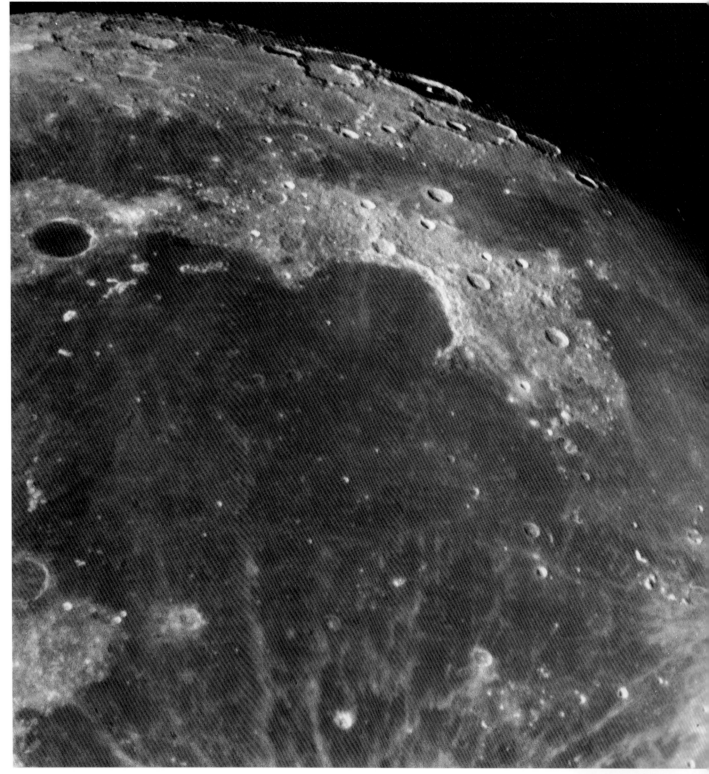

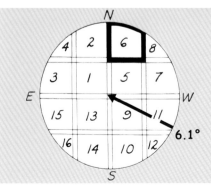

2040 UT 25.12.66. Moon's age 13.8 days. Diameter 64 cm. This was two days before Full Moon. Compare the shape of Plato (b5) with that in Plate 6a. The large crater with a central mountain right on the terminator is Pythagoras (f7).

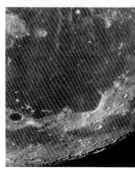

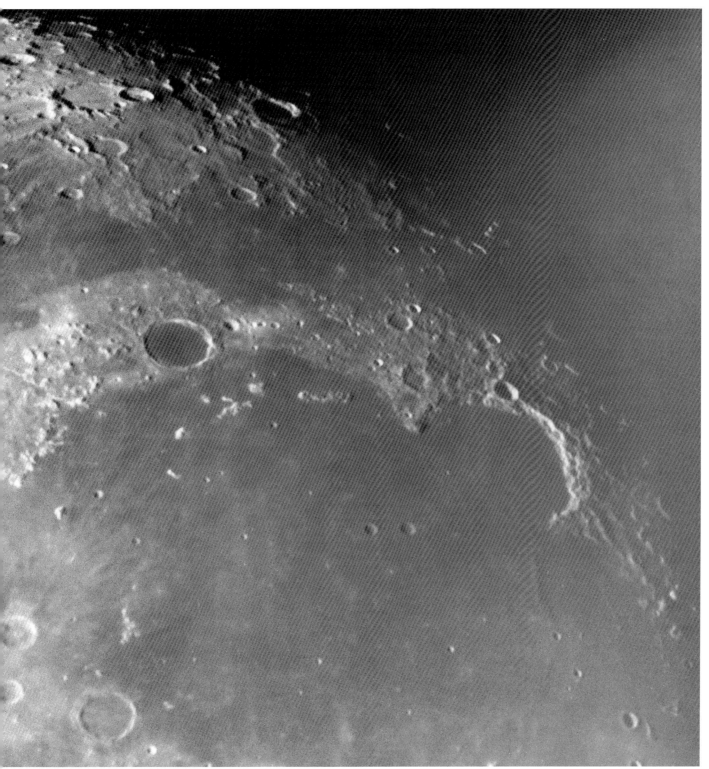

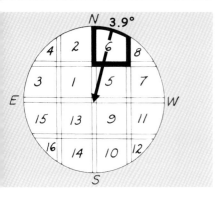

2142 UT 23.11.66. Moon's age 11.3 days. Diameter 64 cm. This is quite a good libration for the north polar regions. The top left-hand (NE) part of this photograph extends beyond Map 6; it is shown on Map 2.

Plate 6d

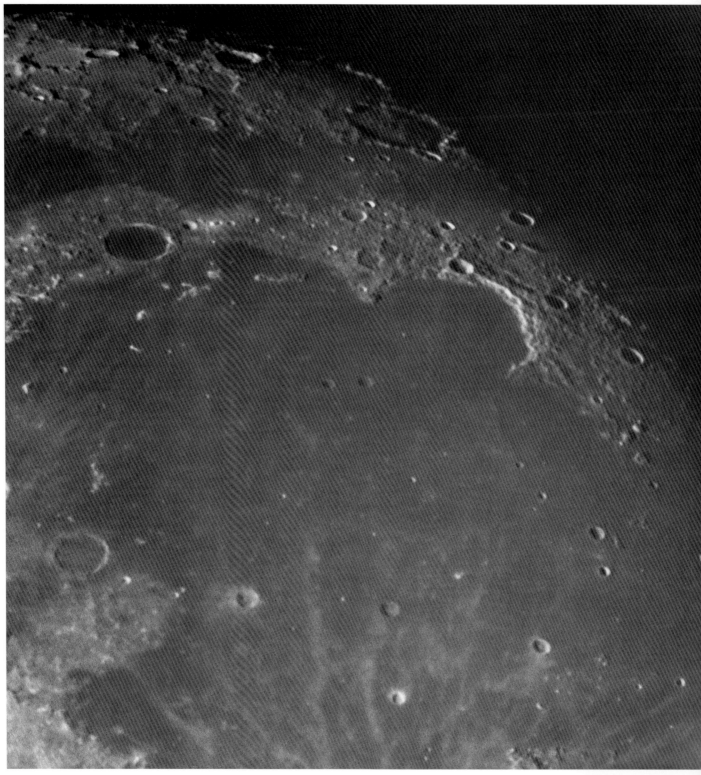

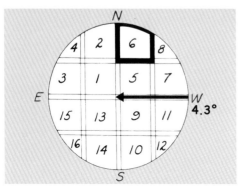

2232 UT 23.12.66. Moon's age 11.8 days. Diameter 64 cm. Compare this with Plate 6c, and note how the different librations cause the shapes of the craters to alter.

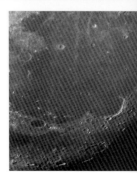

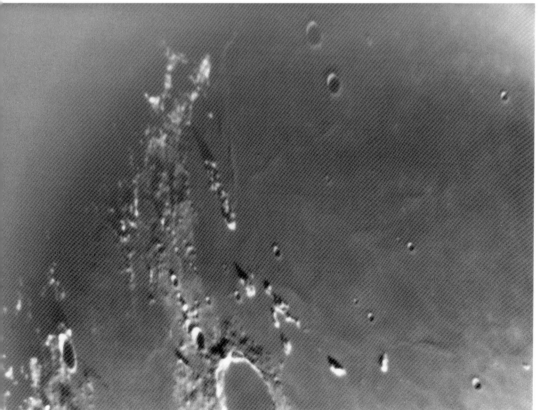

The Sinus Iridum soon after sunrise. 1923 UT 21.03.67. Moon's age 10.6 days. Diameter 91 cm approx. Sinus Iridum lies in Map **6** e4, Cape Heraclides sometimes taking on the appearance of a "Moon Maiden".

The Montes Recti and surrounding country. 2029 UT 20.03.67. Moon's age 9.7 days. Diameter 91 cm approx. The Montes Recti, which lies in Map **6** d5, is a typical isolated mountain range. Its highest peaks are about 1800 m high.

Map 7

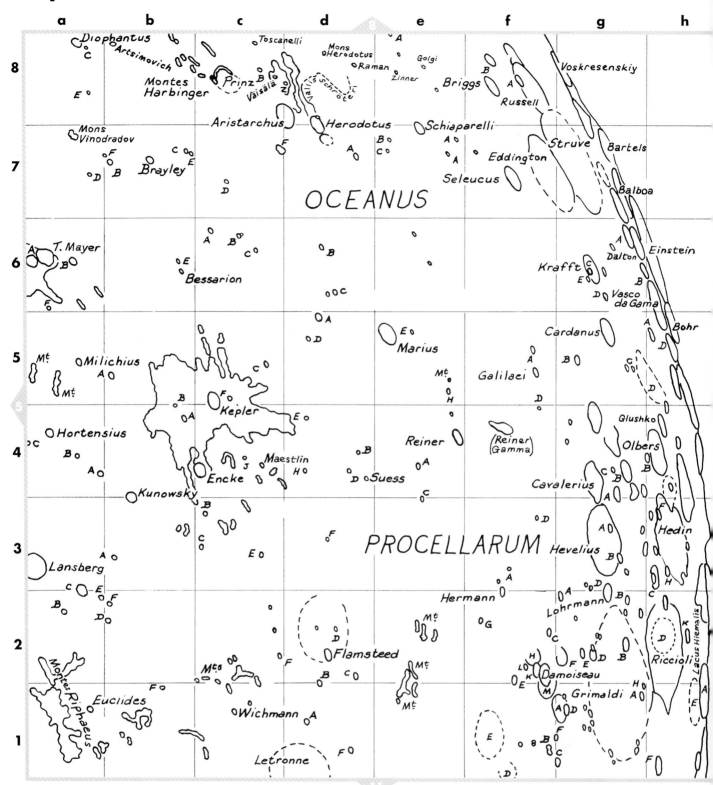

Crater Diameters

Briggs.............................. 37 km (f8)
Diophantus......................... 18 km (a8)
Marius .... 41 km (e5)
Hevelius............................ 106 km (g3)
Euclides .... 11 km (a1)

This map has been prepared from Plates 7a and 7e. Plates 7b, 7c and 7d show same area under different lighting. On Plate 7e there is an insert of the crater Ein (h6), which is only exposed to view when the libration is extremely favourable.

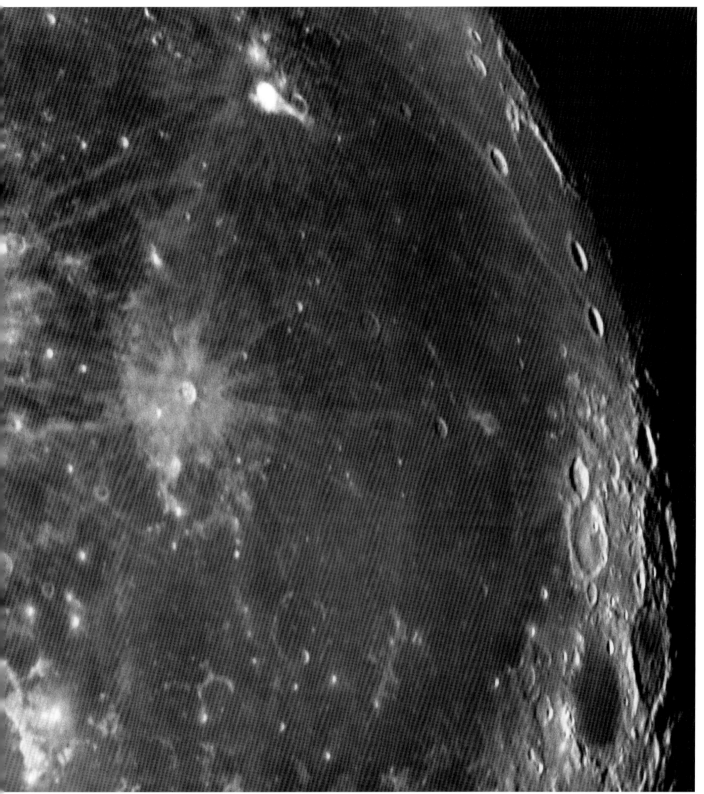

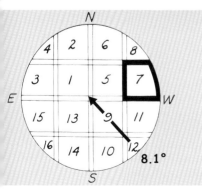

2310 UT 24.01.67. Moon's age 14.2 days. Diameter 64 cm. A comparatively favourable libration here has brought the "limb craters" into view much earlier than usual. Compare the dark floor of Grimaldi (g1) with the extreme brightness of Aristarchus (c8).

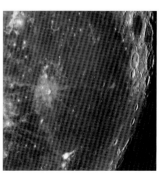

Plate 7b

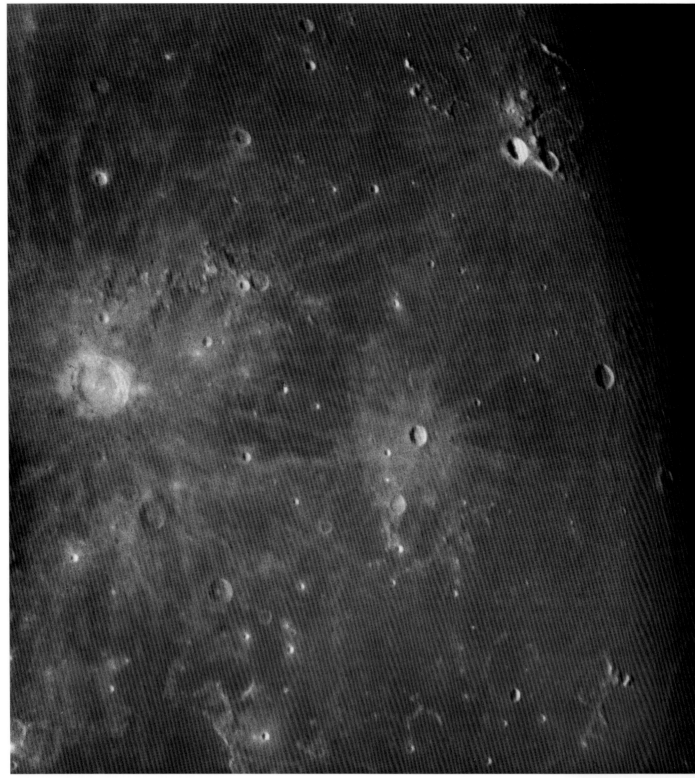

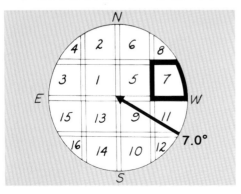

1944 UT 02.02.66. Moon's age 12.1 days. Diameter 64 cm. This photograph extends further to the East than the others in this set. Note Schröter's Valley (d8) and the low ridges near the terminator.

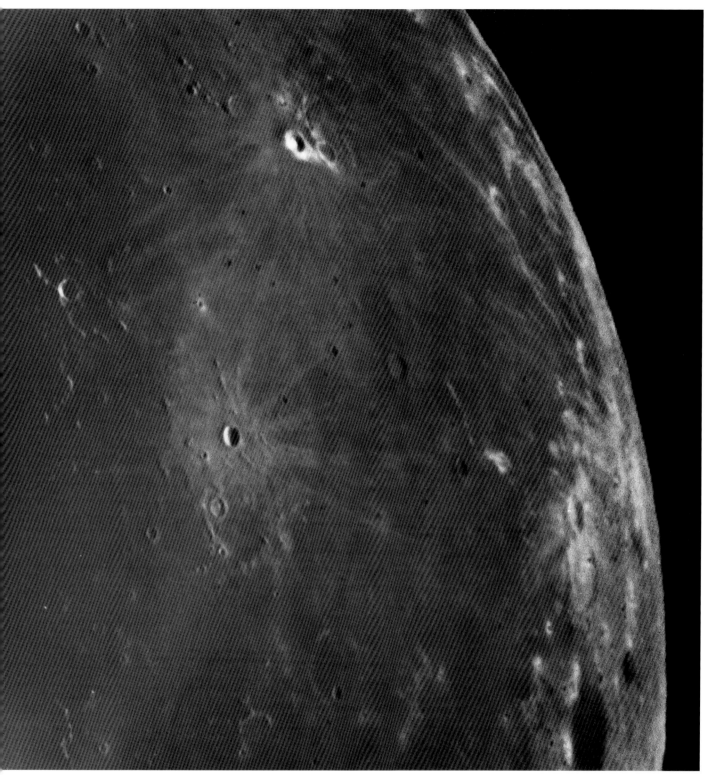

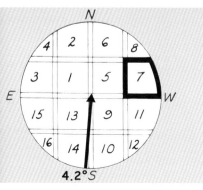

4.2°S

0655 UT 07.12.66. Moon's age 24.6 days. Diameter 64 cm. Compare this with Plate 7a, particularly near the limb.

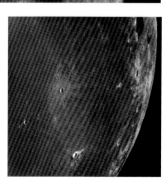

Plate 7d

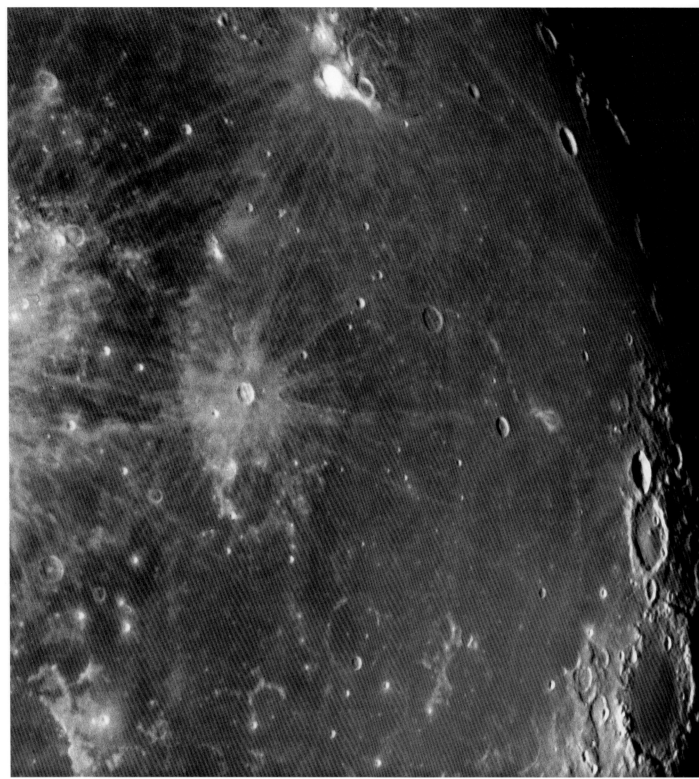

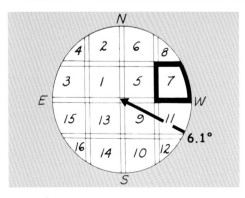

2036 UT 25.12.66. Moon's age 13.8 days. Diameter 64 cm. Note the "St. Andrew's Cross" marking on the Western wall of Grimaldi (g1).

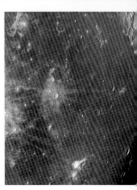

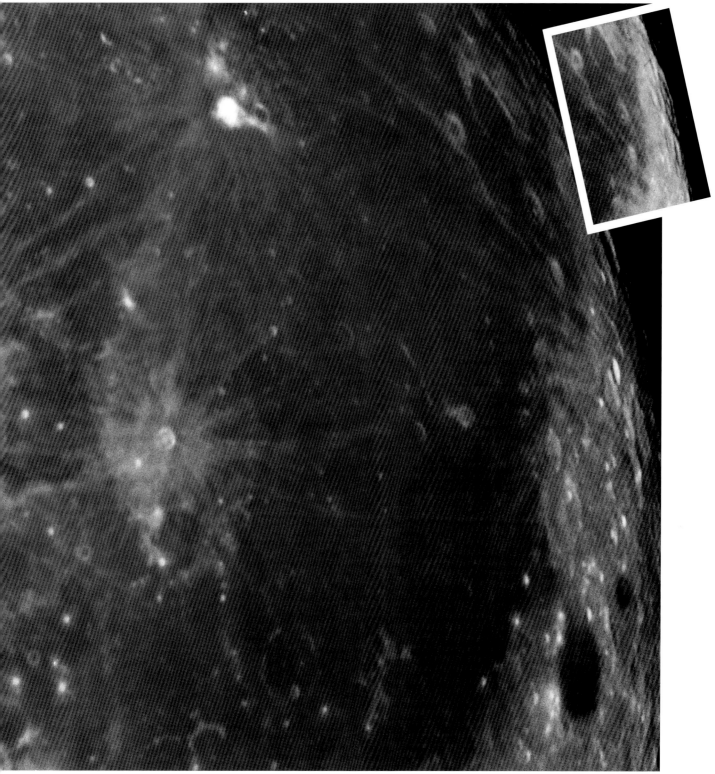

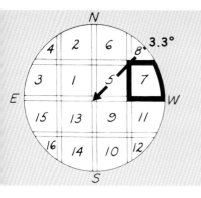

2219 UT 28.10.66. Moon's age 14.7 days. Diameter 64 cm. This was taken 12 hours before Full Moon. Einstein (h6) is just beyond the terminator. Vasco da Gama (h6) is showing plainly. The insert shows Einstein, taken at 2132 UT on 08.11.65, with the author's six-inch reflector. This crater may be well seen on only one or two nights in the average year.

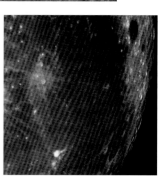

Map 8

Crater Diameters

Pythagoras	130 km	(a8)
Gerard	90 km	(e5)
Seleucus	43 km	(g1)
Wollaston	10 km	(d3)
Euler	28 km	(a2)

This map has been prepared from Plates 8a and 8c. The extreme NE corner c[...] from Plate 6a. Plates 8b and 8d show the same area under different lighting. Pla[...] shows larger scale photographs of the Vallis Schröteri area (e2), Rümker (d5), the Bands in Aristarchus (d2).

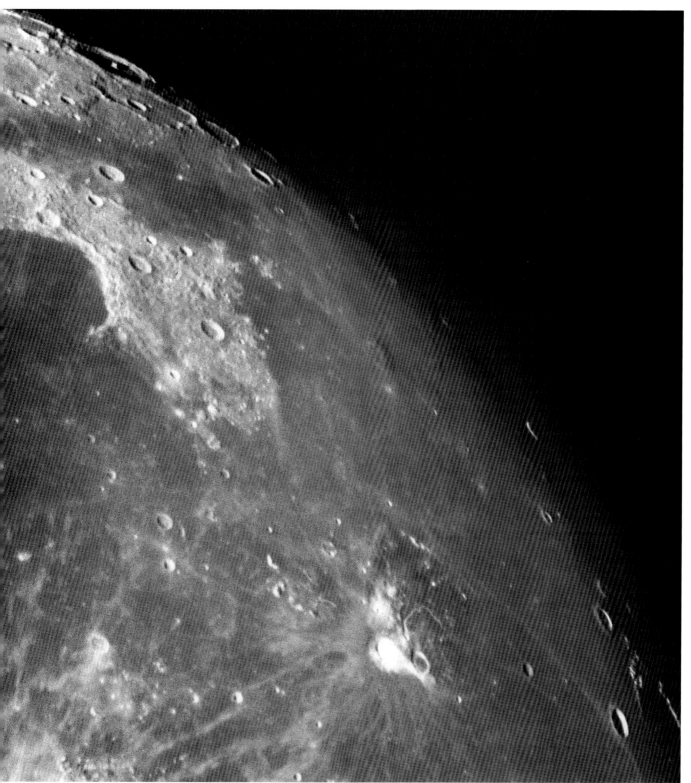

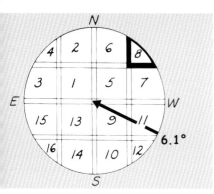

2040 UT 25.12.66. Moon's age 13.8 days. Diameter 64 cm. Pythagoras (a8) is on the terminator near the top. Rümker (d5) looks more like a mound than a crater.

Plate 8b

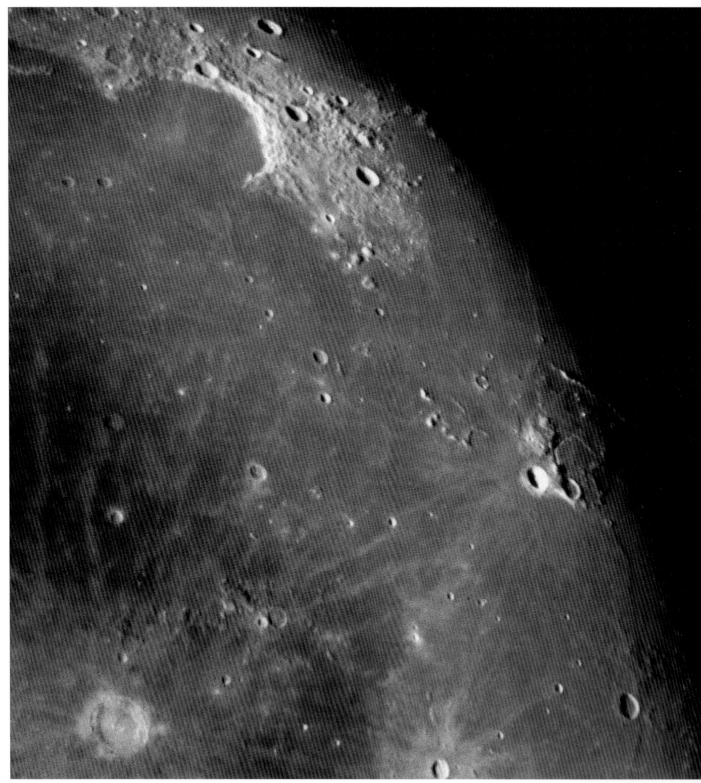

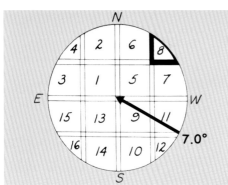

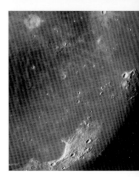

1944 UT 02.02.66. Moon's age 12.1 days. Diameter 64 cm. The Montes Jura (a6) rise to about 6100 m. The Montes Harbinger (c2) are about 2400 m high.

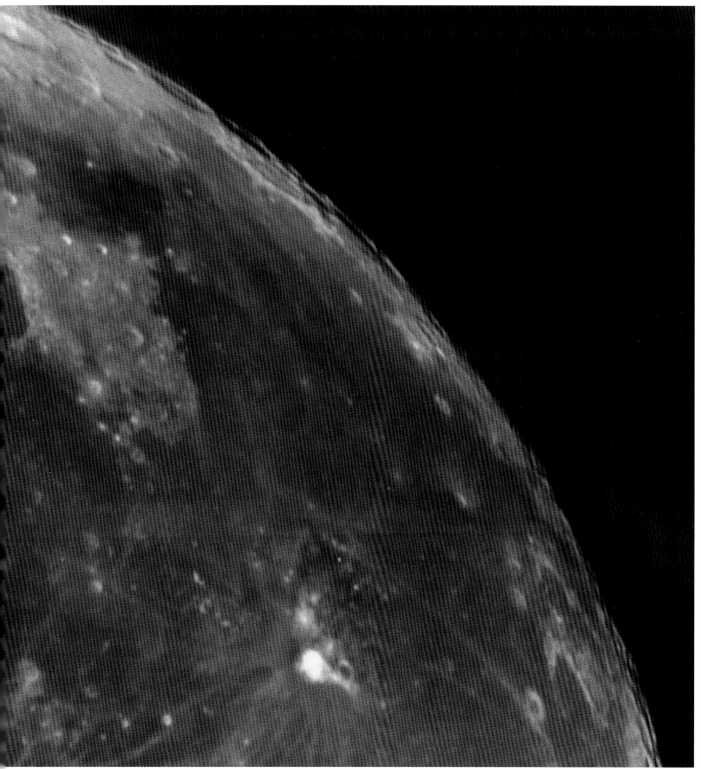

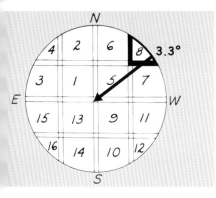

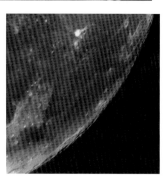

2219 UT 28.10.66. Moon's age 14.7 days. Diameter 64 cm. This was taken 12 hours before Full Moon. The libration is favourable; the craters on the terminator (limb) will not often show up like this.

Plate 8d

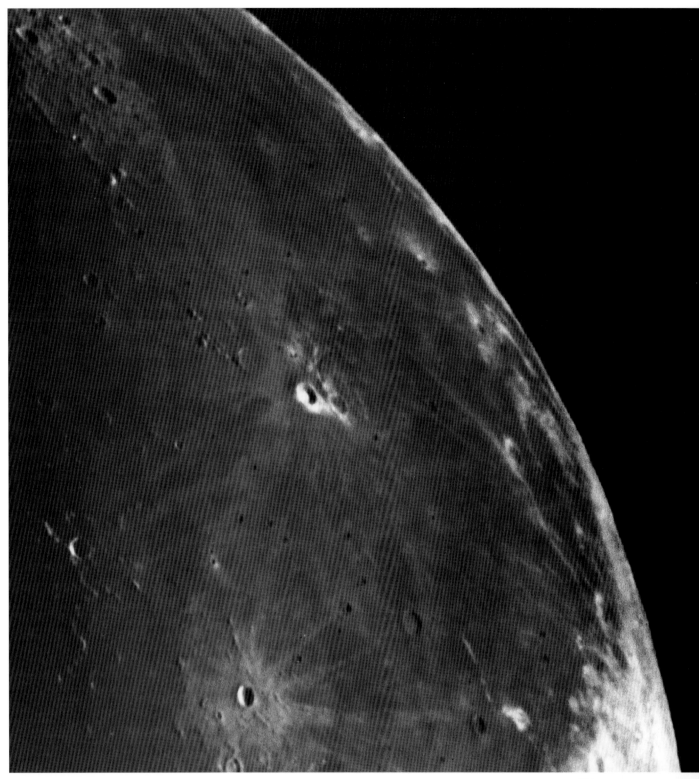

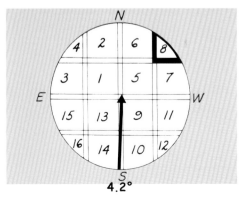

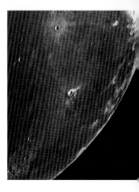

0655 UT 07.12.66. Moon's age 24.6 days. Diameter 64 cm. This is not a favourable libration. Many of the craters shown on Plate 8c are out of sight here, beyond the limb.

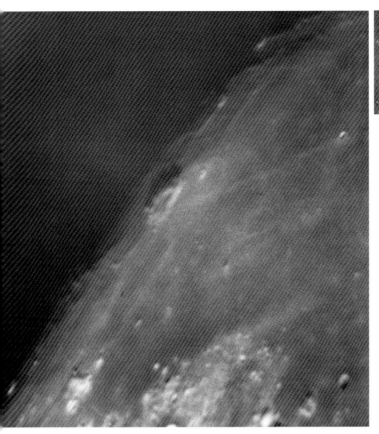

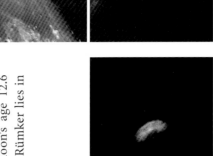

Mons Rümker. 1944 UT 23.3.67. Moon's age 12.6 days. Diameter 89 cm approx. Mons Rümker lies in Map **8** d5.

2110 UT 20.05.67.
Moon's age 11.3 days.

1952 UT 21.04.67.
Moon's age 11.9 days.

2150 UT 23.04.67.
Moon's age 14.0 days.

Bands in Aristarchus (d2). All these photographs had exposures much less than normal, so that Aristarchus was properly exposed; compare them with the Plate to the left. All are to the same scale, the Moon's diameter being 94 cm approx.

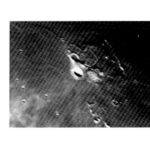

Vallis Schröteri, Aristarchus and Herodotus. 2033 UT 21.04.67. Moon's age 12.0 days. Diameter 94 cm approx. Vallis Schröteri lies on Map **8** e2.

Map 9

Crater Diameters

Flamsteed	21 km	(h8)
Gambart	25 km	(c8)
Bullialdus	61 km	(e3)
Lepaute	16 km	(g2)
Hell	33 km	(b1)

This map has been prepared from Plate 9a. Plates 9b, 9c, 9d and 9e show the area under different lighting. Plate 9f shows larger scale photographs of the Recta area (b3). Plate 9g shows larger scale photographs of the area round Bulli and Kies (e2), and of the Mare Humorum clefts and ridges (f3).

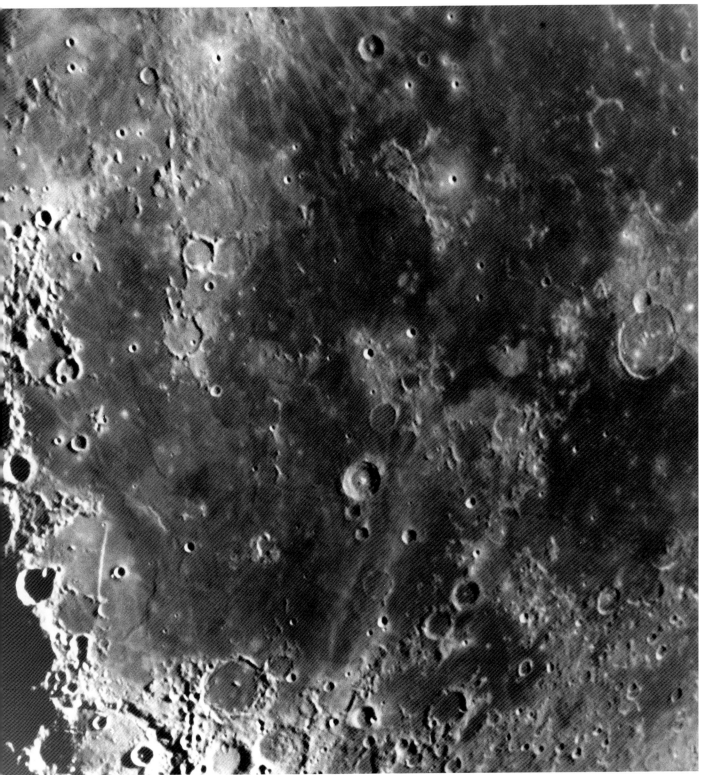

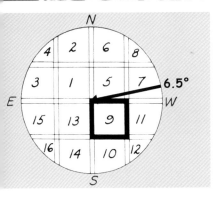

0317 UT 09.08.66. Moon's age 23.8 days. Diameter 64 cm. Note the Rupes Recta or Straight Wall (b3); it is about 110 km long and 250 m to 300 m high, and slopes up towards the West. The Montes Riphaeus (f6) are about 900 m high.

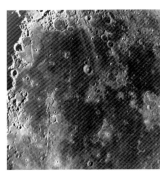

Plate 9b

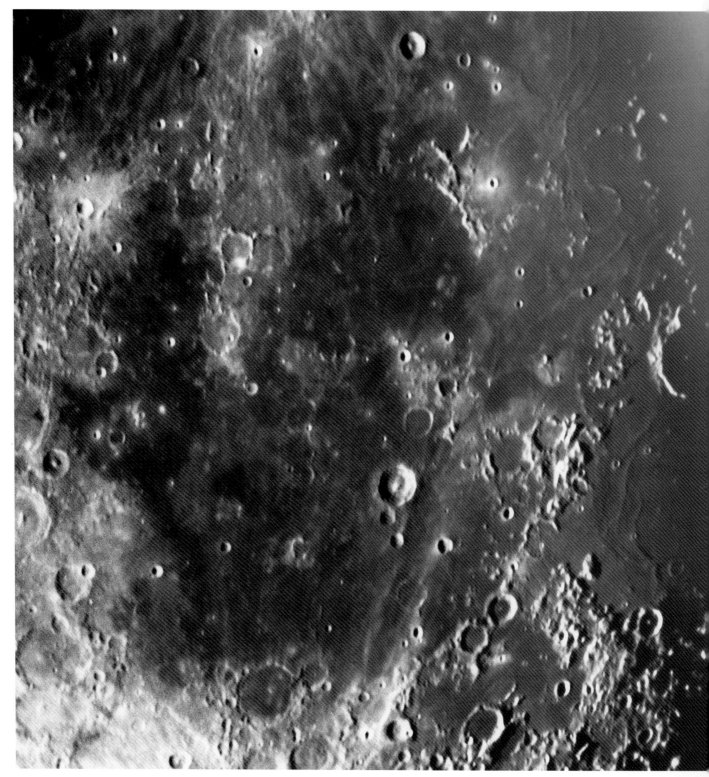

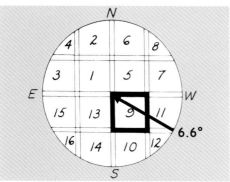

1758 UT 21.01.67. Moon's age 11.0 days. Diameter 64 cm. Note the rilles NW from Campanus (f2) and the ridges running down the right-hand side of the photograph. These are only visible when the lighting is very oblique. (See Plate 9e also.)

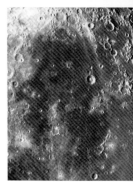

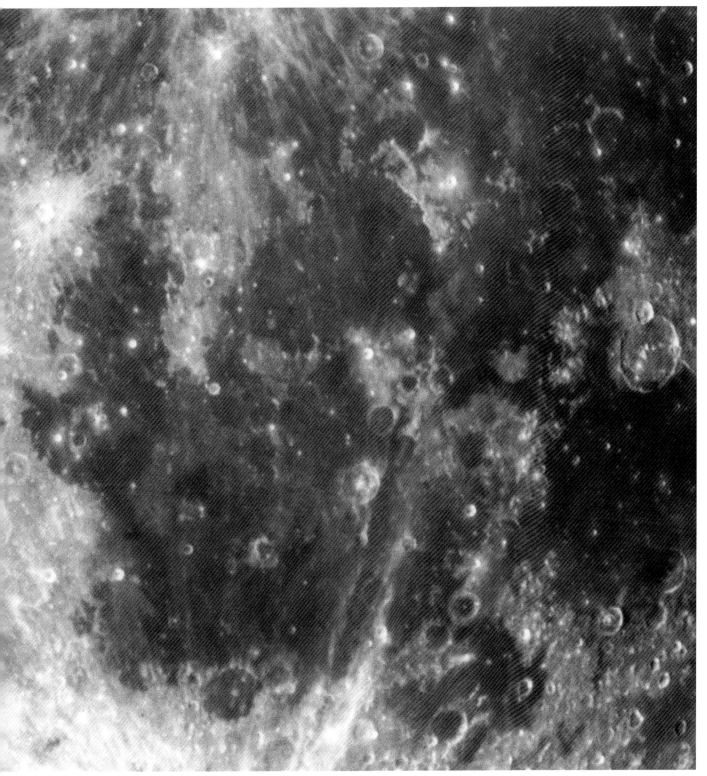

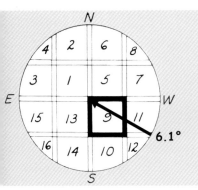

2135 UT 25.12.66. Moon's age 13.8 days. Diameter 64 cm. Compare this with Plates 9b and 9a, which show almost the same area under morning and afternoon lighting.

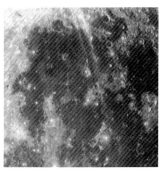

Plate 9d

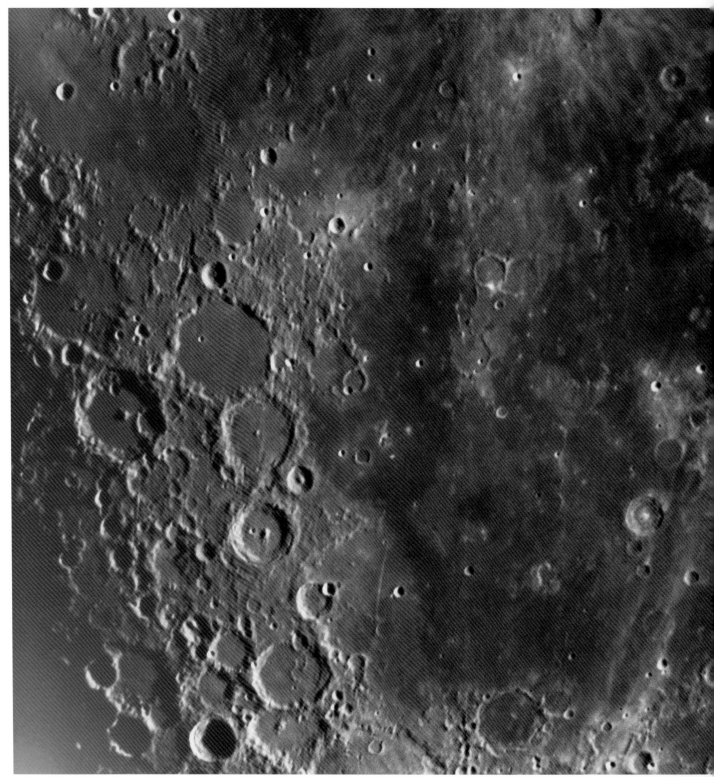

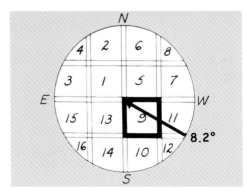

0516 UT 06.10.66. Moon's age 21.4 days. Diameter 64 cm. This photograph extends further to the East than the others in this group and overlaps into sections 1 and 13.

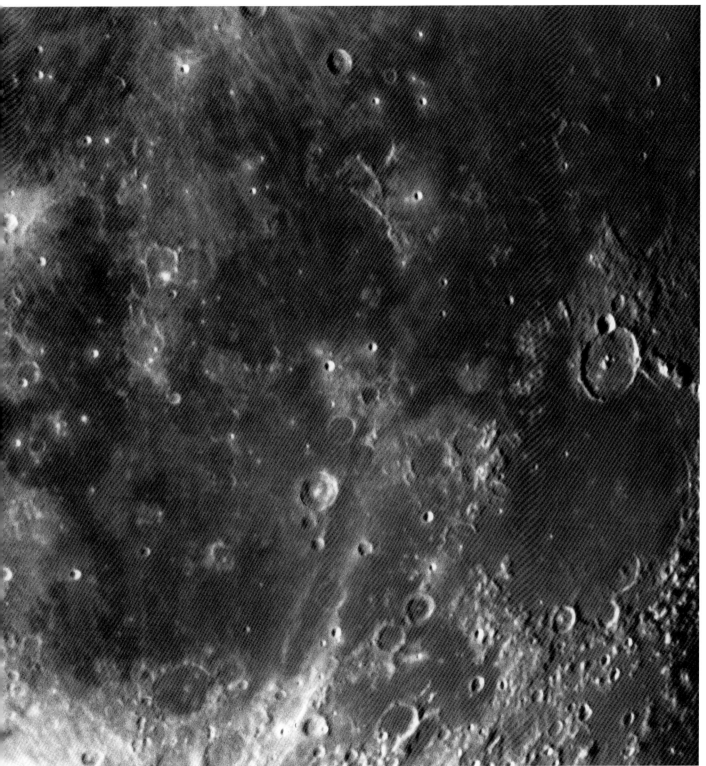

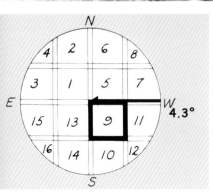

2236 UT 23.12.66. Moon's age 11.8 days. Diameter 64 cm. The Moon here is less than one day older than it is in Plate 9b and yet the rilles and dorsa shown in the latter have virtually disappeared.

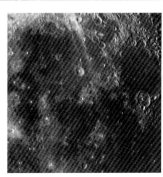

Plate 9f

0317 UT 09.08.66. Moon's age 23.8 days.

1928 UT 19.03.67. Moon's age 8.6 days.

Early morning (left) and late afternoon (right) views of the Rupes Recta or Straight Wall area. See Map **9** b3. The Moon's diameter in both

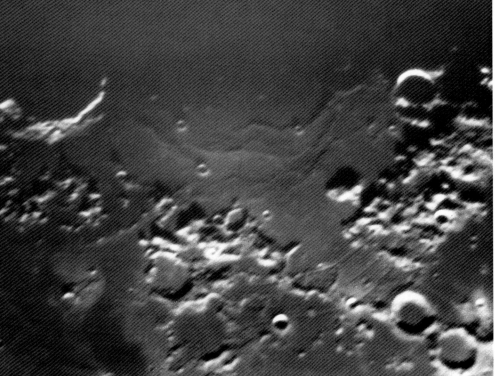

Rilles and dorsa on the Eastern shores of the Mare Humorum. 1932 UT 21.03.67. Moon's age 10.6 days. Diameter 91 cm approx. Campanus and Mercator (top left) lie in Map 9 f2.

Bullialdus, Kies and the Kies dome, and the Hesiodus Cleft. 2056 UT 20.03.67. Moon's age 9.7 days. Diameter 91 cm approx. The Kies dome lies just under one diameter West from Kies, which is shown in Map 9 e2.

Map 10

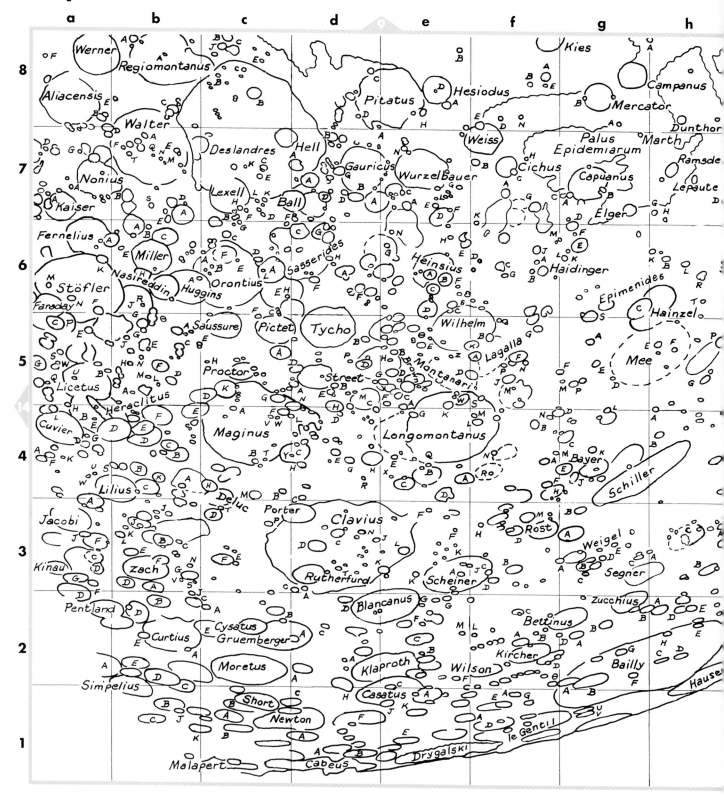

Crater Diameters

Mercator	47 km	(g8)
Werner	70 km	(a8)
Tycho	85 km	(d5)
Bettinus	71 km	(g2)
Moretus	114 km	(c2)

This map has been prepared from Plate 10a. Plates 10b, 10c and 10d show the s
area under different lighting. Plate 10e shows larger scale photographs of the
between Clavius (d3) and Bailly (g2).

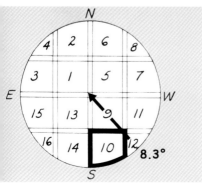

0609 UT 04.11.66. Moon's age 21.0 days. Diameter 64 cm. This is a favourable libration for this area.

Plate 10b

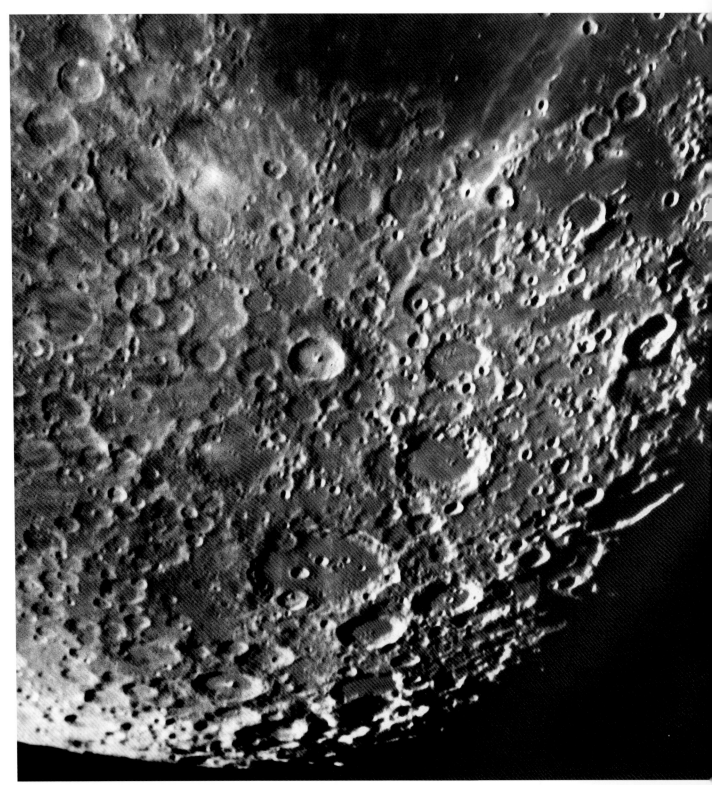

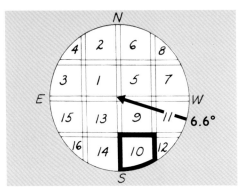

1758 UT 21.01.67. Moon's age 11.0 days. Diameter 64 cm. Compare this with Plate 10a; the libration here is not so favourable.

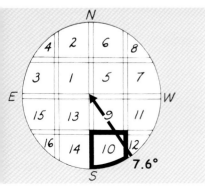

2104 UT 23.02.67. Moon's age 14.5 days. Diameter 64 cm. This was taken about 20 hours before Full Moon. Compare it with Plate 10d, which shows almost the same area.

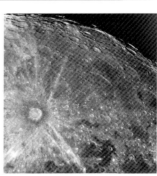

Plate 10d

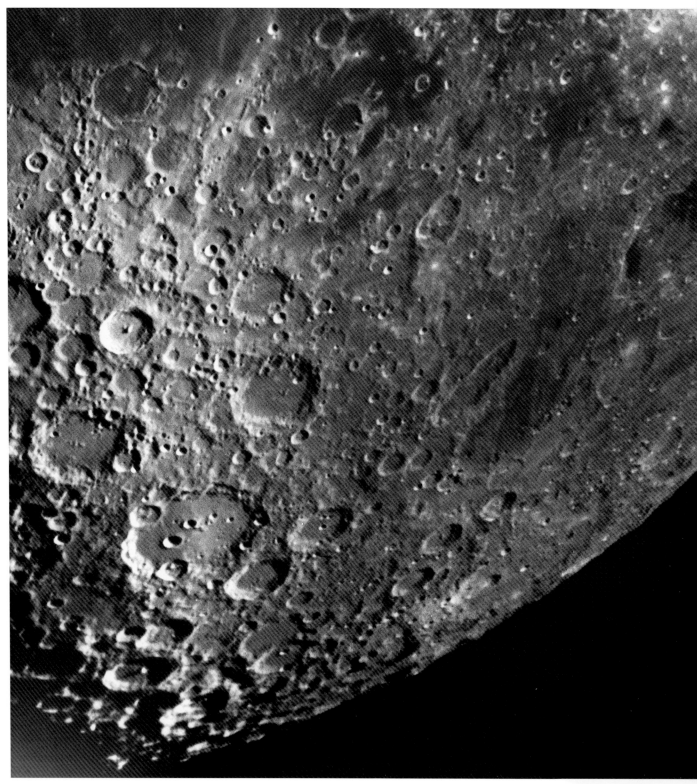

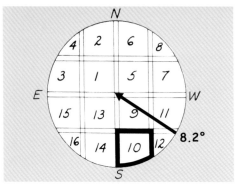

0516 UT 06.10.66. Moon's age 21.4 days. Diameter 64 cm. Compare this with Plates 10a and 10c.

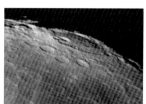

...y. 2259 UT 05.03.66. Moon's age 13.5 days. Diameter 91 cm approx. Bailly lies in Map **10** g2.

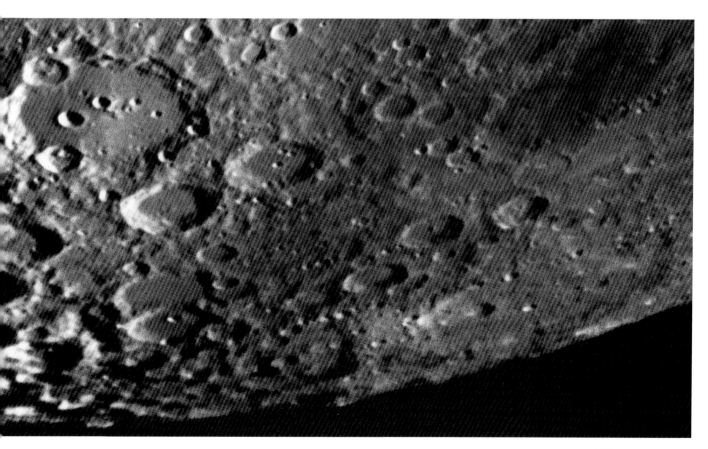

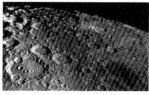

...us to Bailly. 0516 UT 06.10.66. Moon's age 21.4 days. Diameter 91 cm approx. This is an enlarge-
...: of part of Plate 10d. Compare the Bailly area here with the same region in the top photograph.

Map 11

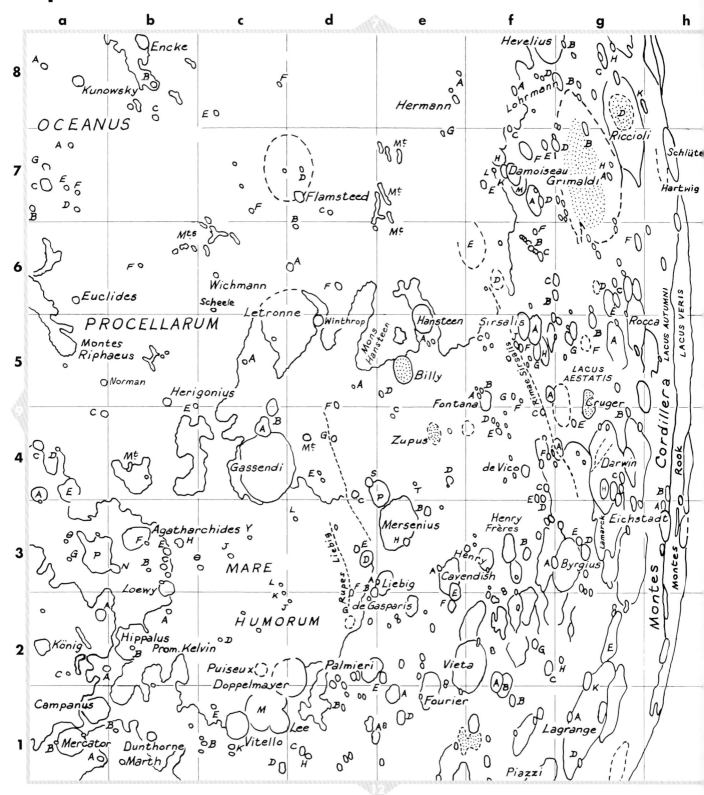

a   **b**   **c**   **d**   **e**   **f**   **g**   **h**

Encke

Kunowsky

OCEANUS

Hevelius

Hermann

Lohrmann

Riccioli

Schlüter

Damoiseau

Grimaldi

Hartwig

Flamsteed

Euclides

PROCELLARUM

Wichmann

Scheele

Letronne

Winthrop

Hansteen

Mons Hansteen

Sirsalis

Rocca

Montes Riphaeus

Norman

Herigonius

Billy

Fontana

LACUS AESTATIS

Cruger

Gassendi

Zupus

de Vico

Darwin

Henry Frères

Eichstadt

Lamarck

Mersenius

Rupes Liebig

Henry

Cavendish

Byrgius

Agatharchides

MARE

Rimae Sirsalis

LACUS AUTUMNI

LACUS VERIS

Cordillera

Montes Rook

Montes

Loewy

Liebig

de Gasparis

HUMORUM

König

Hippalus

Prom. Kelvin

Puiseux

Doppelmayer

Palmieri

Vieta

Lagrange

Campanus

Lee

Fourier

Mercator

Dunthorne

Marth

Vitello

Piazzi

Crater Diameters

Damoiseau .... 37 km (f7)
Kunowsky .... 18 km (a8)
Billy .... 46 km (e5)
Vieta .... 87 km (f2)
Dunthorne .... 16 km (b1)

This map has been prepared from Plates 11a and 11c. Plates 11b and 11d show same area under different lighting. Plate 11e shows larger scale photograph Darwin, Rimae Sirsalis and Grimaldi (g4 to g7) and of the Mare Humorum Gassendi (c3).

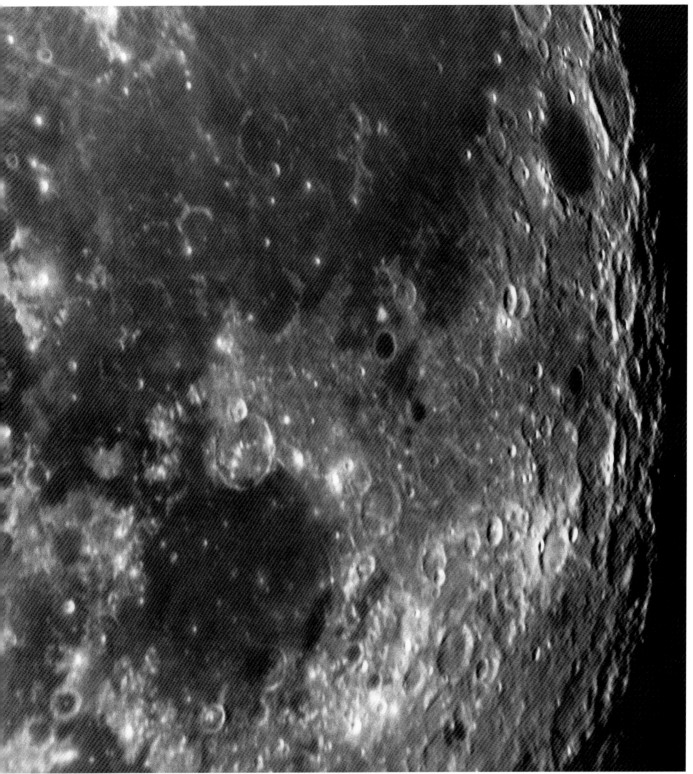

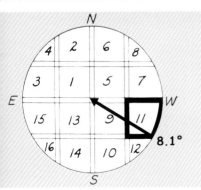

2310 UT 24.01.67. Moon's age 14.2 days. Diameter 64 cm. This is a fairly favourable libration for this area. Byrgius A (f3) may be seen as a "ray centre" on Plates 11b and 11c.

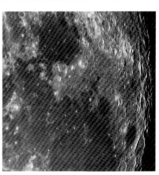

Plate 11b

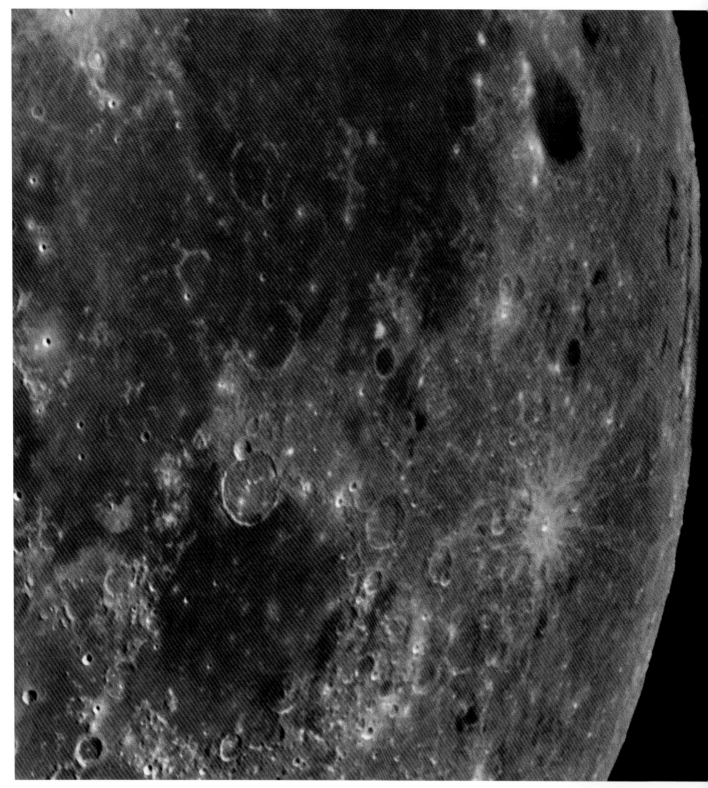

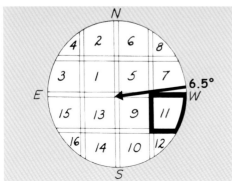

0317 UT 09.08.66. Moon's age 23.8 days. Diameter 64 cm. Compare this with Plate 11a. The "ray centre" (top right) is Byrgius A (f3).

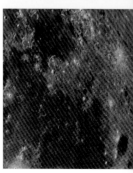

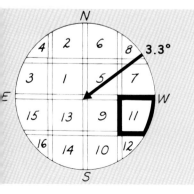

2229 UT 28.10.66. Moon's age 14.7 days. Diameter 64 cm. This was taken about 12 hours before Full Moon. Compare the detail on the limb here with that on Plate 11b. Although the libration in Plate 11b is better than it is here, there is virtually nothing to be seen on the limb, since the lighting is not suitable.

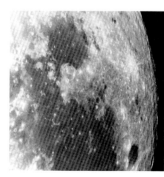

Plate 11d

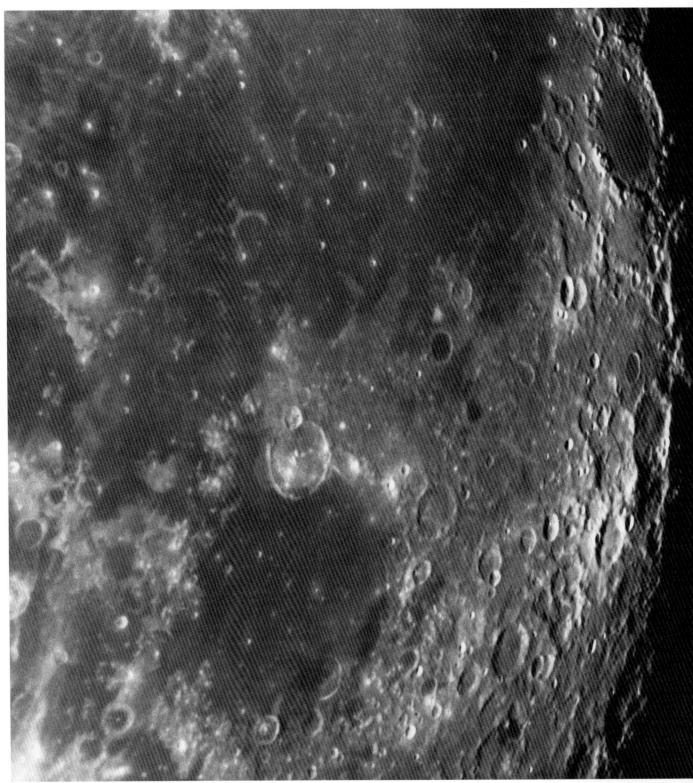

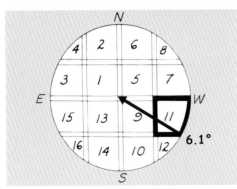

2036 UT 25.12.66. Moon's age 13.8 days. Diameter 64 cm. Rimae Sirsalis (f5) will probably show up more clearly here if the reader turns the page 90° in a clockwise direction.

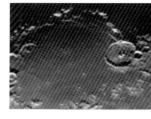
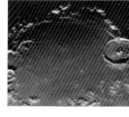

Above: The Mare Humorum and Gassendi (c4). 2106 UT 20.04.67. Moon's age 11.0 days. Diameter 94 cm approx.

Left: Darwin, Rimae Sirsalis and Grimaldi. 2036 UT 25.12.66. Moon's age 13.8 days. Diameter 91 cm approx. This is an enlargement of part of Plate 11d. Note the "St. Andrew's Cross" marking on the NW wall of Grimaldi (g7).

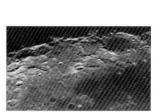

Map 12

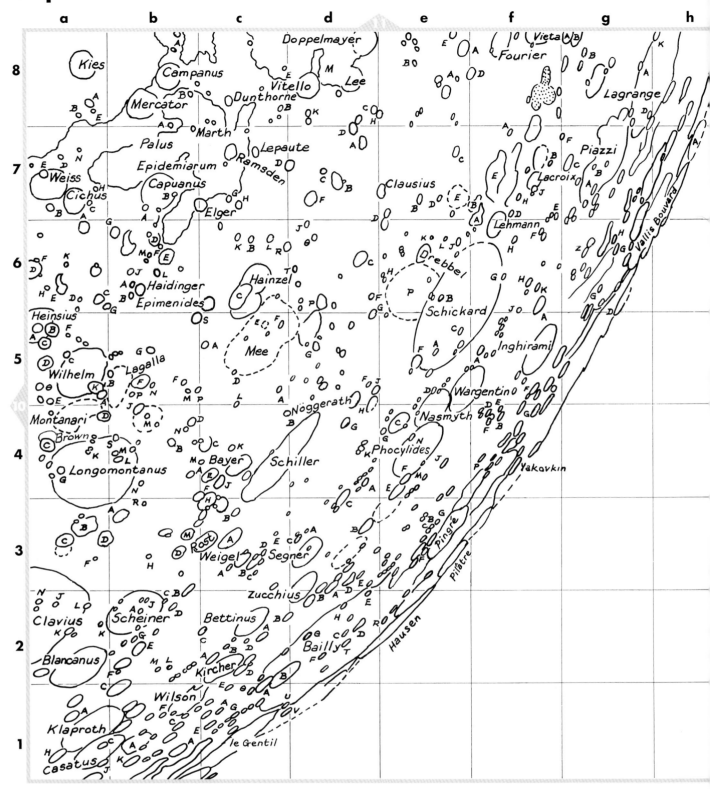

This map has been prepared from Plate 12a. Plates 12b, 12c and 12d show the area under different lighting. Plate 12e shows larger scale photographs of Bailly and the area around Wargentin (e5).

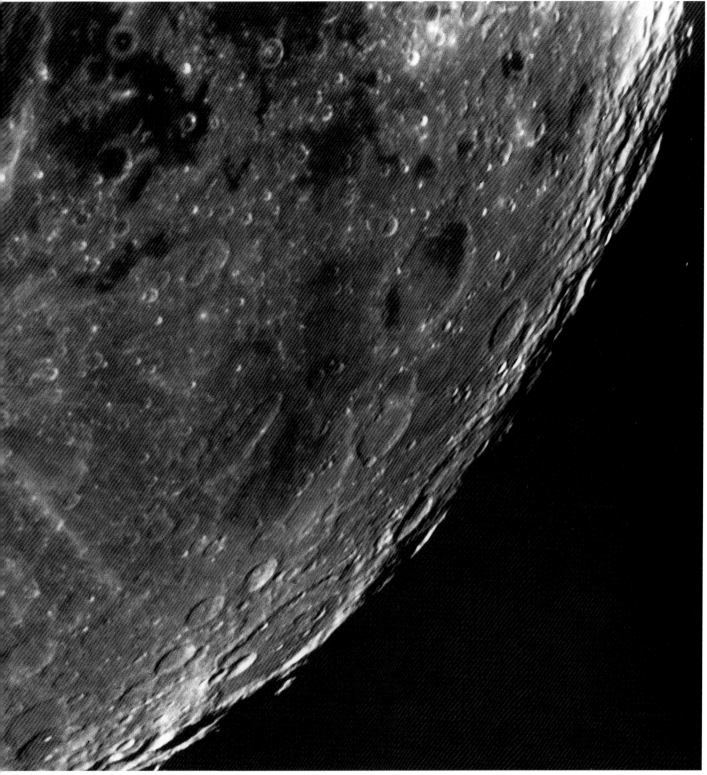

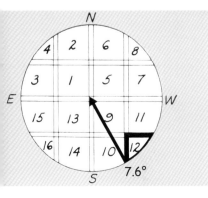

2104 UT 23.02.67. Moon's age 14.5 days. Diameter 64 cm. This is a good libration for this area. Bailly (d2) will not often be seen as well as this.

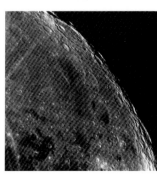

Plate 12b

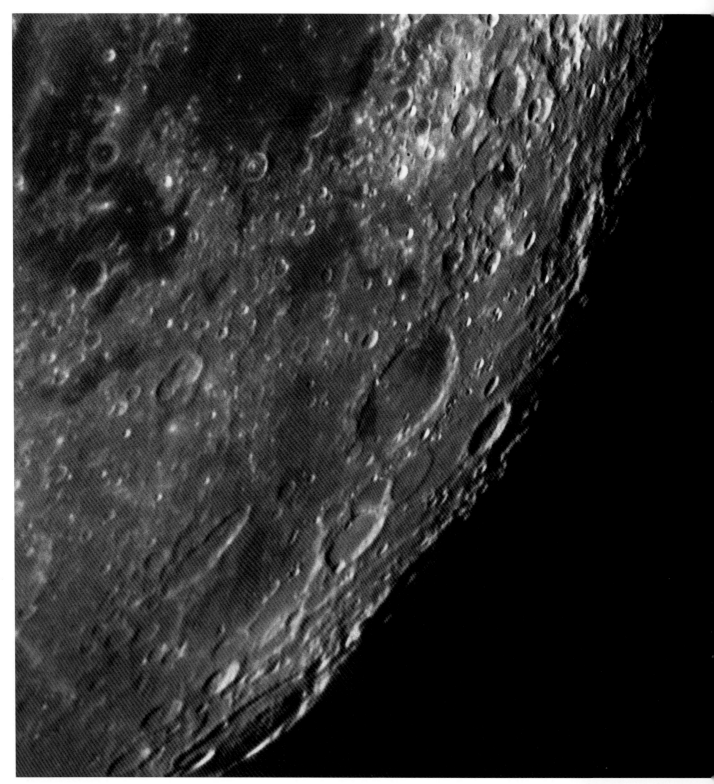

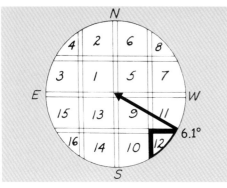

2036 UT 25.12.66. Moon's age 13.8 days. Diameter 64 cm. Note that Wargentin (e5) has been filled in to form a plateau crater.

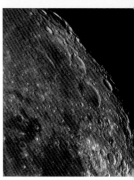

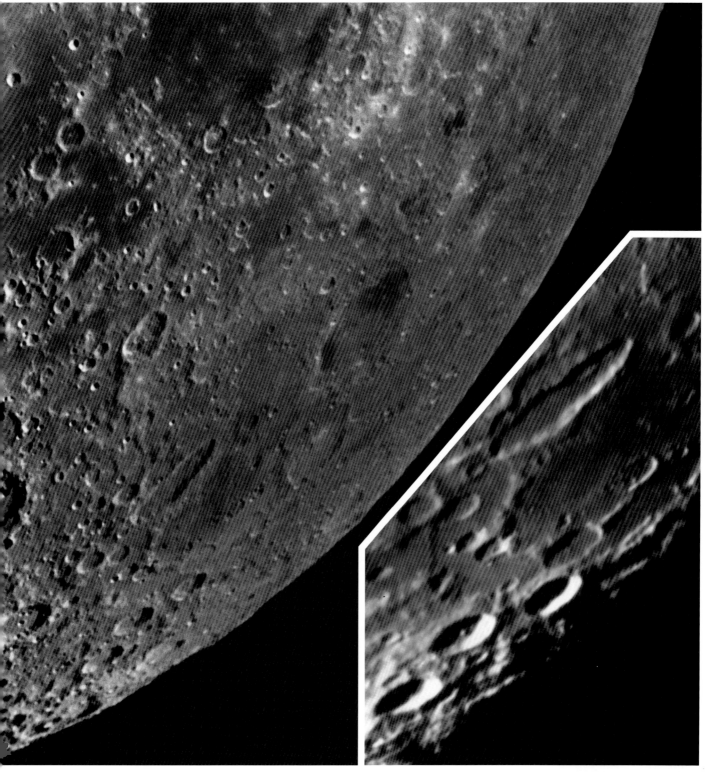

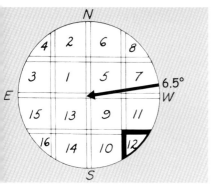

0317 UT 09.08.66. Moon's age 23.8 days. Diameter 64 cm. Compare this with Plate 12a, particularly near Bailly (d2). *Inset:* Kircher (c2) to Schiller (c4). 1958 UT 21.04.67. Moon's age 11.9 days. Diameter 94 cm approx.

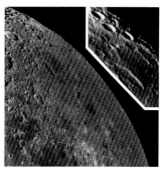

Plate 12d

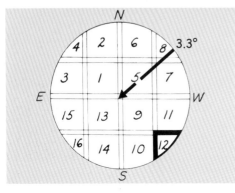

2229 UT 28.10.66. Moon's age 14.7 days. Diameter 64 cm. This was taken 12 hours before Full Moon. Compare it with Plates 12a and 12c.

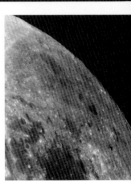

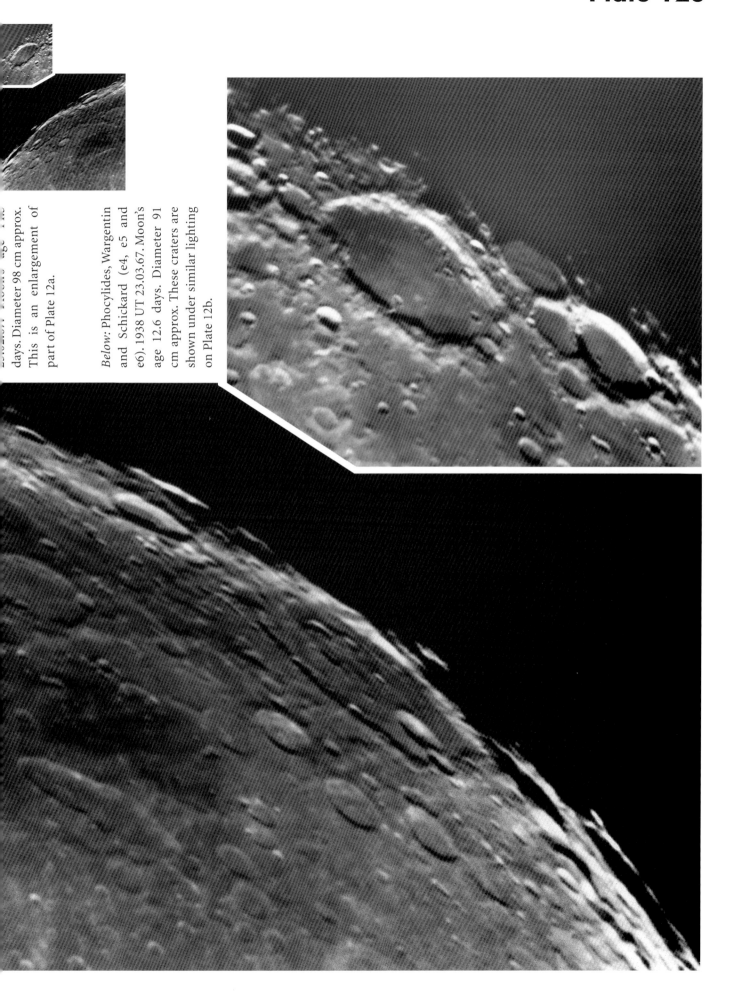

days. Diameter 98 cm approx. This is an enlargement of part of Plate 12a.

Below: Phocylides, Wargentin and Schickard (e4, e5 and e6). 1938 UT 23.03.67. Moon's age 12.6 days. Diameter 91 cm approx. These craters are shown under similar lighting on Plate 12b.

Map 13

Crater Diameters

Mösting	25 km	(h8)
Sabine	30 km	(b8)
Argelander	34 km	(e4)
Werner	70 km	(f2)
Rothmann	42 km	(b1)

This map has been prepared from Plate 13a. Plates 13b, 13c and 13d show the area under different lighting. Plate 13e shows larger scale photographs of Abe (d3) and the area round Walter (g1). Plate 13f shows the area from Ptolemaeus (g Hell (h1) on a larger scale. Plate 13g shows the Ptolemaeus–Alphonsus–Arz chain on a larger scale.

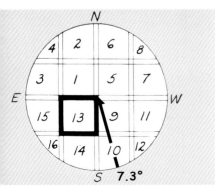

2046 UT 28.04.66. Moon's age 7.9 days. Diameter 64 cm. Note the small crater Regiomontanus A (g2) which lies on the summit of a mountain.

Plate 13b

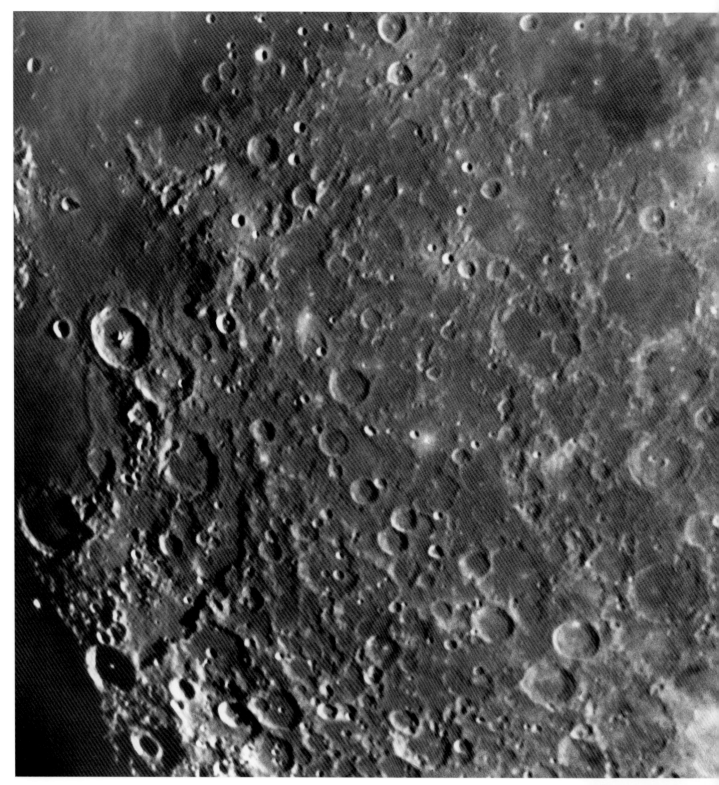

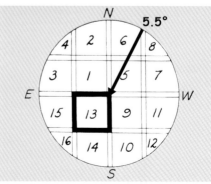

0215 UT 06.08.66. Moon's age 18.9 days. Diameter 64 cm. The Rupes Altai (b2) form an escarpment which is about 1800 m high generally; individual peaks may rise as much again.

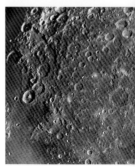

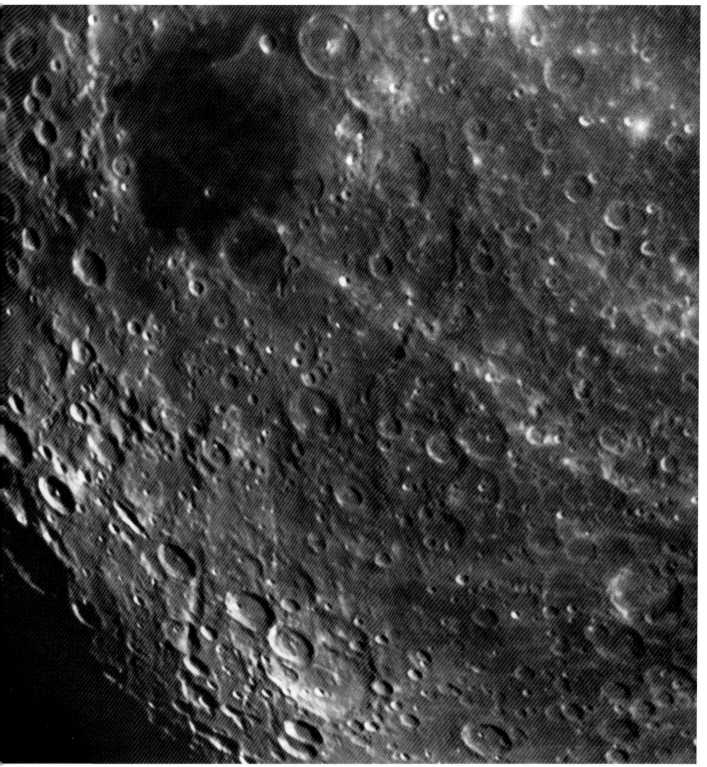

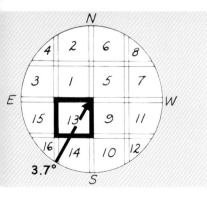

2334 UT 26.02.67. Moon's age 17.6 days. Diameter 64 cm. This photograph extends further to the South and East than the other members of this group. At the bottom it overlaps into the areas covered by Maps 14 and 16.

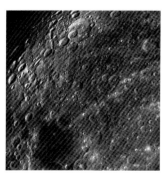

Plate 13d

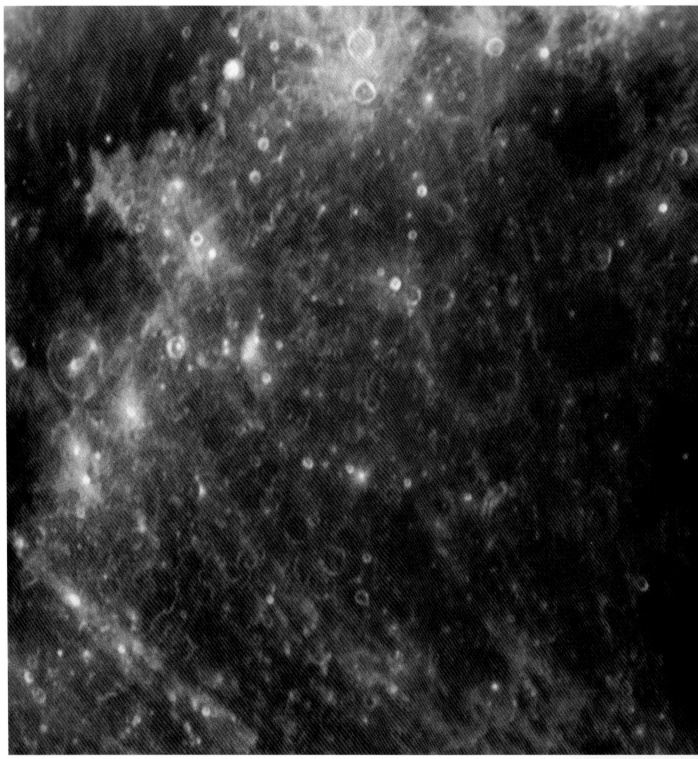

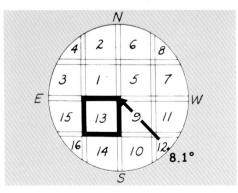

2314 UT 24.01.67. Moon's age 14.2 days. Diameter 64 cm. This was taken just over one day before Full Moon. Compare it with Plates 13a and 13b which show almost exactly the same area. Almost all the bright rays radiate from Tycho (Map **10** d5).

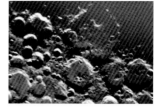

The area round Walter (g1). 1950 UT 17.04.67. Moon's age 7.9 days. Diameter 94 cm approx. Note the mountain-top crater Regiomontanus A (g2) and the dark bands in Stöfler (top left).

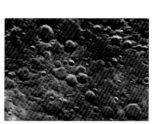

The area round Abenezra (d3). 2008 UT 16.05.67. Moon's age 7.2 days. Diameter 94 cm approx. Note the radial dark bands in Abenezra C.

Plate 13f

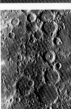

Ptolemaeus (g6) to Hell (h1). 1959 UT 19.03.67. Moon's age 8.6 days. Diameter 91 cm.

Ptolemaeus, Alphonsus and Arzachel (g6, g5 and g4) about two days after sunrise. 2050 UT 20.03.67. Moon's age 9.7 days. Diameter 91 cm approx. Note the dark patches in Alphonsus (g5).

Ptolemaeus, Alphonsus and Arzachel (g6, g5 and g4) soon after sunrise. 1951 UT 17.04.67. Moon's age 7.9 days. Diameter 91 cm approx. The diameter of Albategnius C (f5) is 6 km.

Map 14

b   c   d   e   f   g   h

Crater Diameters

Walter	....	140 km (f7)
Stiborius	....	44 km (a7)
Cuvier	....	75 km (e4)
Clavius	....	225 km (g3)
Manzinus	....	98 km (d2)

This map has been prepared from Plate 14a. Plates 14b, 14c and 14d show the area under different lighting. Plate 14e shows larger scale photographs of the round Maurolycus (d6) and Stöfler (e6), and Clavius (g3).

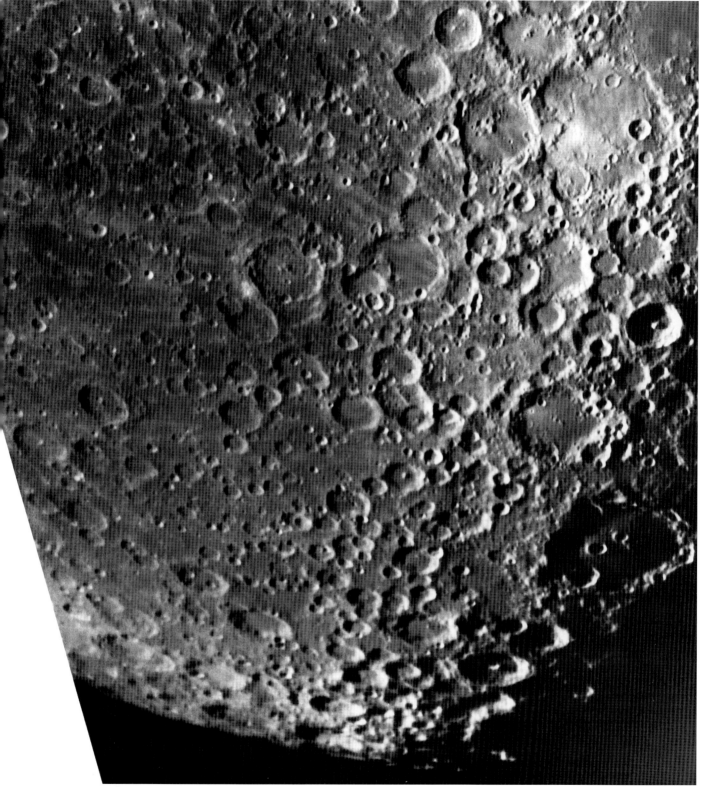

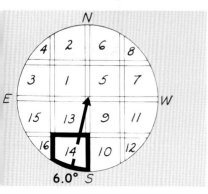

2122 UT 28.05.66. Moon's age 8.4 days. Diameter 64 cm. This is quite a good libration for the area at the top of the photograph.

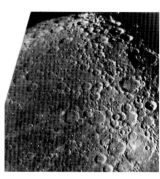

Plate 14b

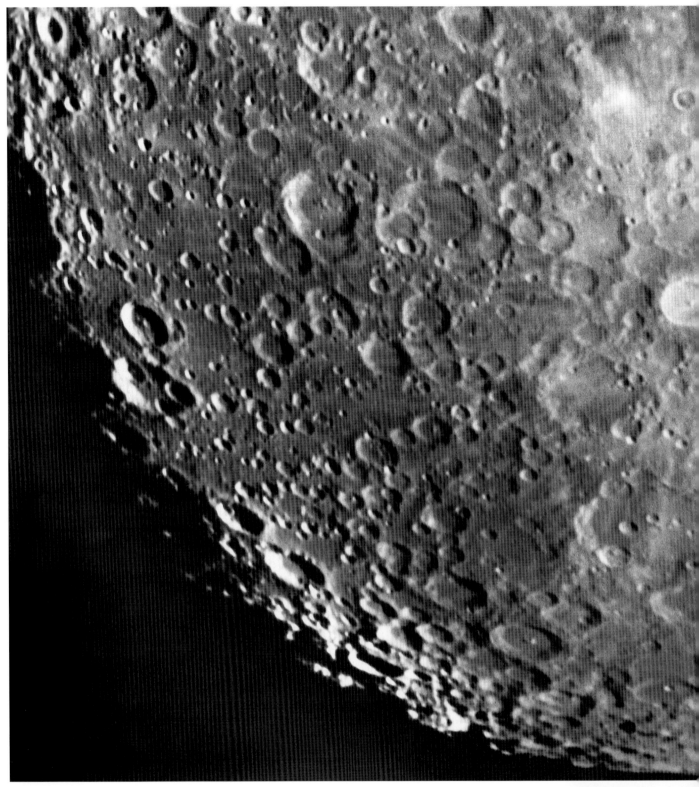

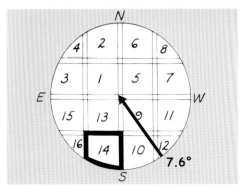

2336 UT 01.12.66. Moon's age 19.4 days. Diameter 64 cm. Compare this with Plates 14a and 14c, which show almost exactly the same area.

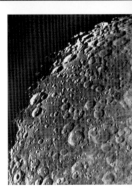

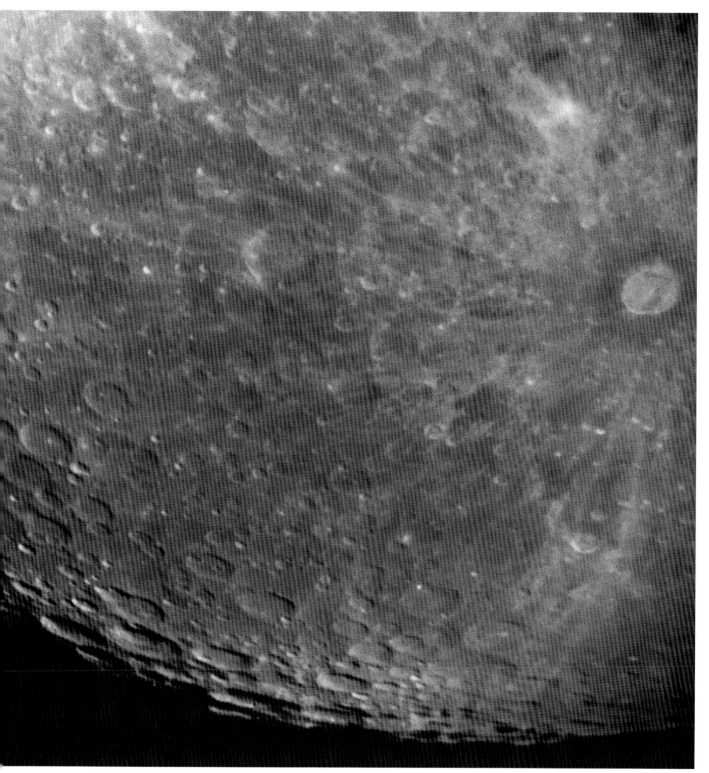

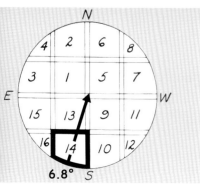

2326 UT 06.02.66. Moon's age 16.3 days. Diameter 64 cm. This was taken just over one day after Full Moon and the rays from Tycho are very prominent. Compare this with Plates 14a and 14b, which show almost exactly the same area.

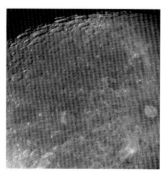

Plate 14d

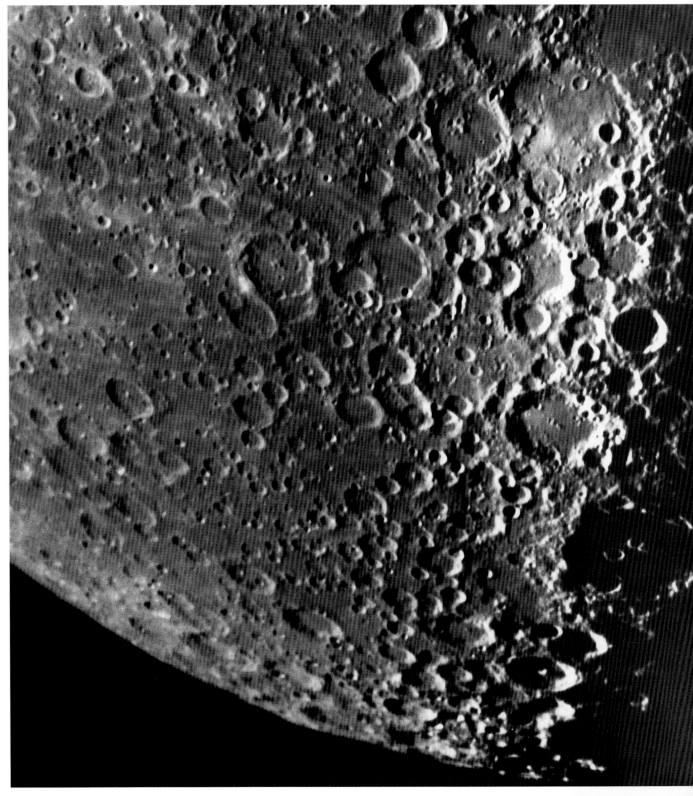

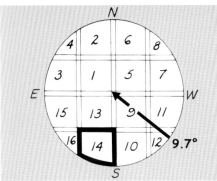

1959 UT 19.03.67. Moon's age 8.6 days. Diameter 64 cm. Compare this with Plate 14a. The Moon's age is very nearly the same in both cases but the librations are quite different.

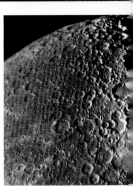

The area around Clavius (g3). 2046 UT 20.03.67. Moon's age 9.7 days. Diameter 91 cm approx. The small crater Clavius CB (g3) is 7 km in diameter.

The area around Maurolycus (d6) and Stöfler (e6). 2021 UT 16.05.67. Moon's age 7.3 days. Diameter 94 cm approx. Note that both Maurolycus (d6) and Ideler (c5) are overlapping smaller craters.

Map 15

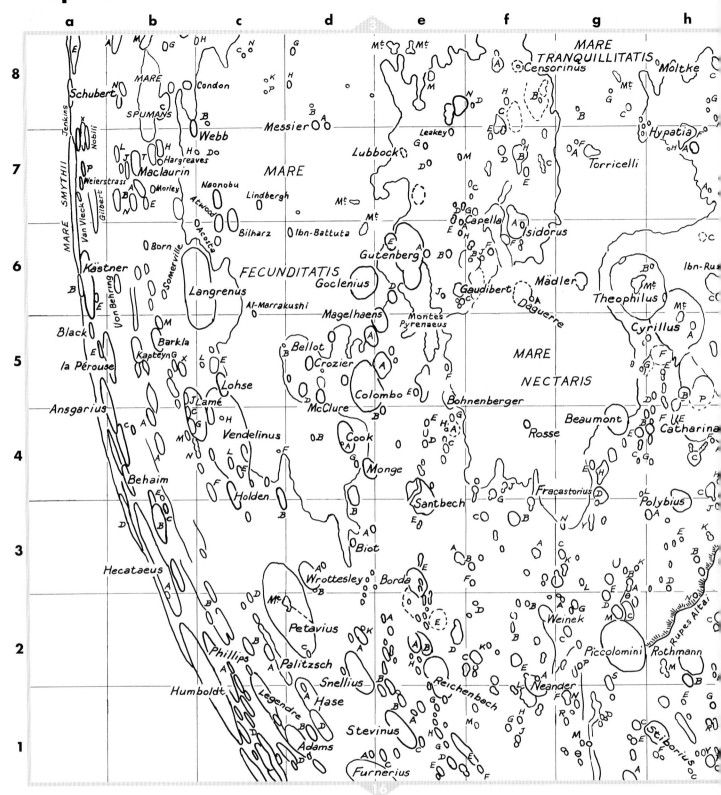

Crater Diameters

Moltke	7 km	(h8)
Webb	22 km	(b8)
Theophilus	100 km	(g6)
Piccolomini	88 km	(g2)
Stevinus	75 km	(e1)

This map has been prepared from Plates 15a and 15d. Plates 15b, 15c and 15e the same area under different lighting. Plate 15e also shows a larger scale photo of the area round Humboldt (c2).

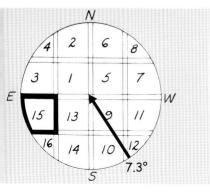

2152 UT 27.04. 66. Moon's age 6.9 days. Diameter 64 cm. Compare this with the other members of this group, which show the detail near the limb much more clearly.

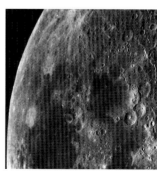

Plate 15b

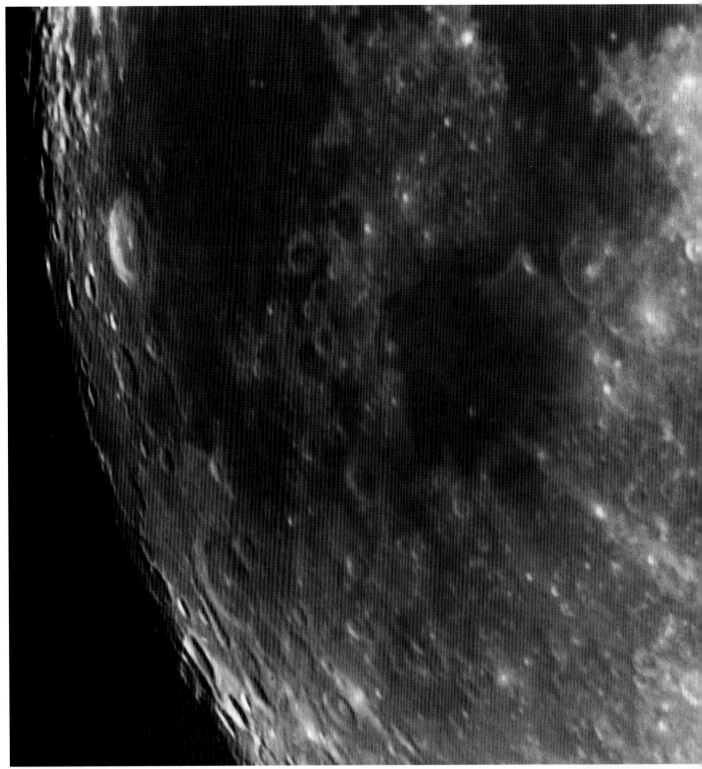

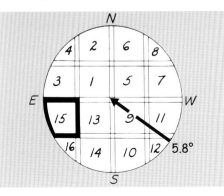

2125 UT 28.11.66. Moon's age 16.3 days. Diameter 64 cm. This was taken about 19 hours after Full Moon.

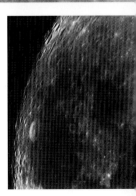

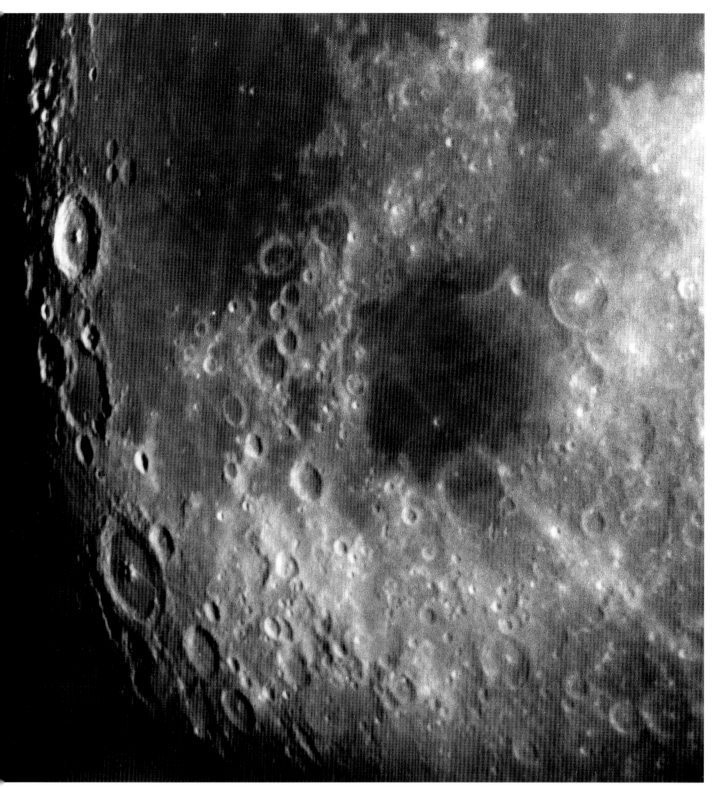

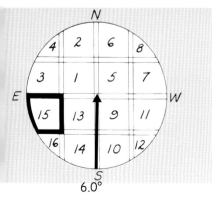

2228 UT 08.01.66. Moon's age 17.0 days. Diameter 64 cm. Petavius (d2) and Langrenus (c6) are very prominent. Compare this with Plate 15b which shows almost exactly the same area.

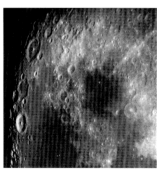

Plate 15d

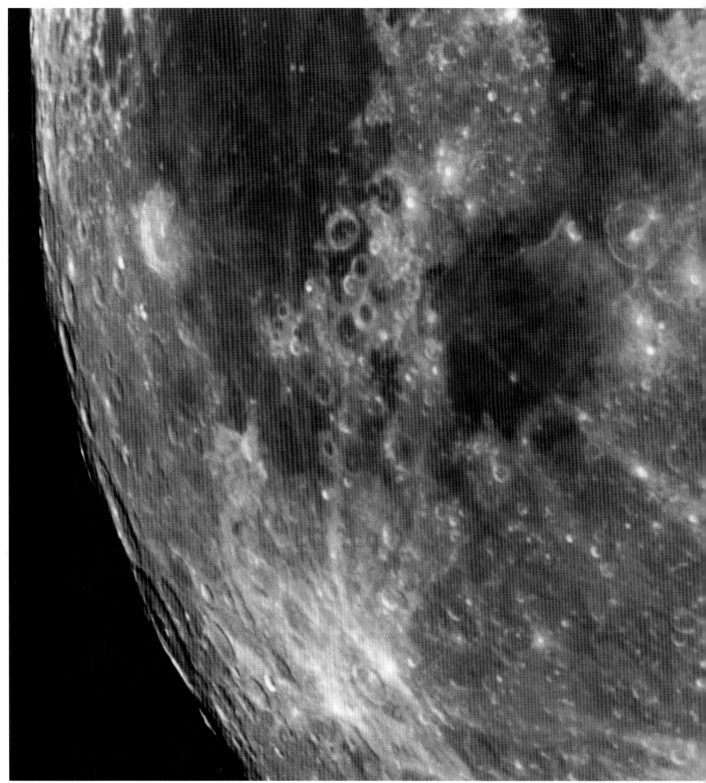

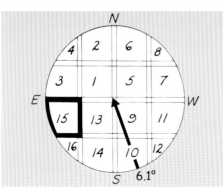

2300 UT 24.02.67. Moon's age 15.6 days. Diameter 64 cm. This was taken about 5 hours after Full Moon. The libration is not favourable, but even so the craters on the limb near Humboldt (c1) will not often be seen like this.

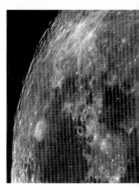

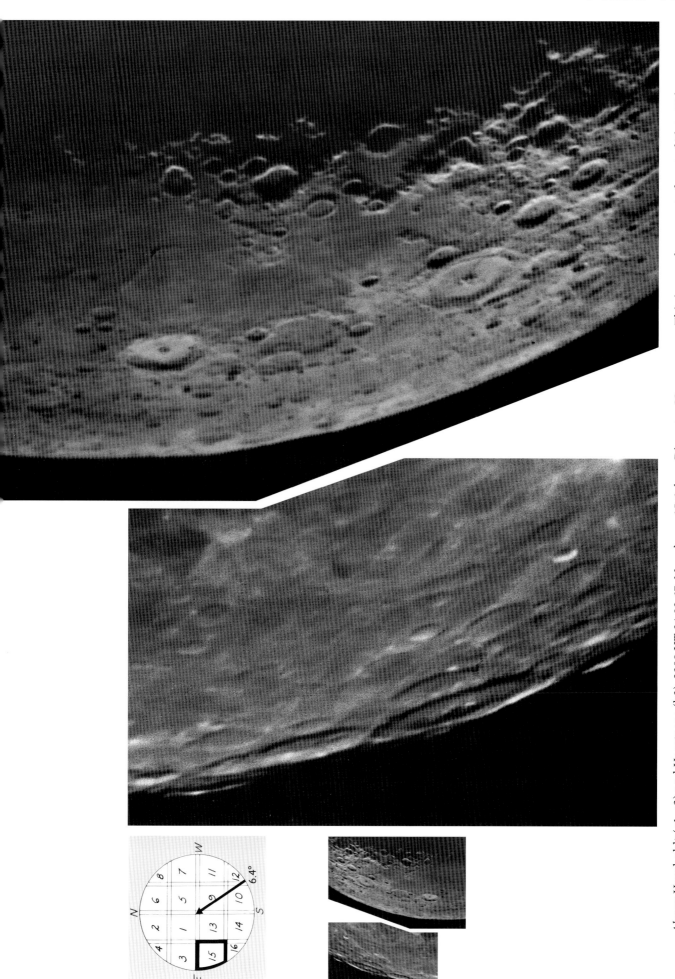

Above: Humboldt (c1, c2) and Hecataeus (b3). 2300 UT 24.02.67. Moon's age 15.6 days. Diameter 99 cm approx. This is an enlargement of part of Plate 15d.

Right: 2033 UT 23.05.66. Moon's age 3.4 days. Diameter 64 cm. Compare this with Plate 15d. The librations are very similar.

Map 16

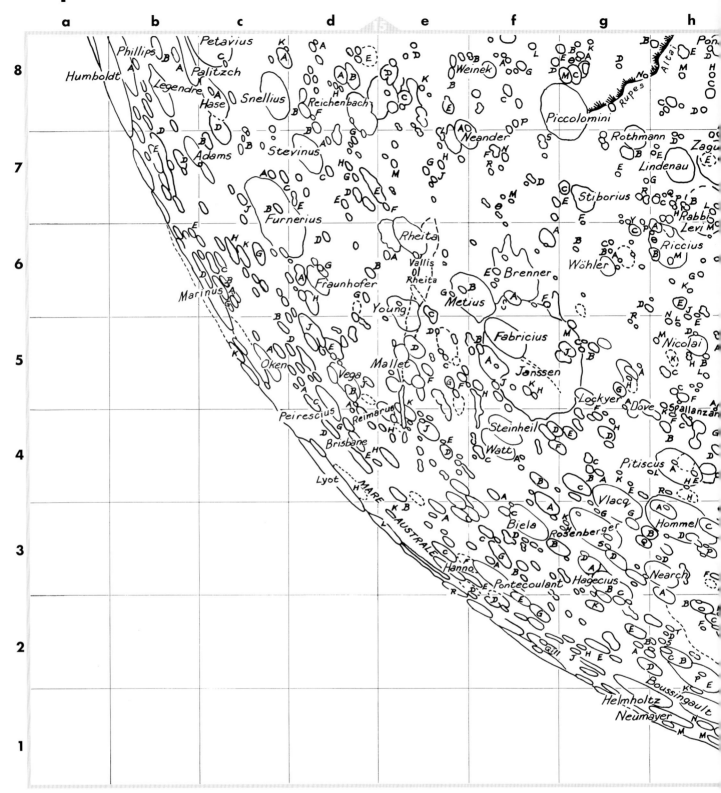

This map has been prepared from Plates 16a and 16c. Plates 16b, 16d and 16e
the same area under different lighting. Plate 16f shows a larger scale photogra
the Mare Australe area (d4).

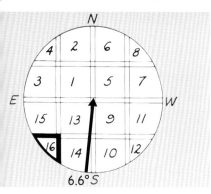

2109 UT 23.06.66. Moon's age 5.0 days. Diameter 64 cm. Compare this with Plate 16c which shows almost exactly the same area.

Plate 16b

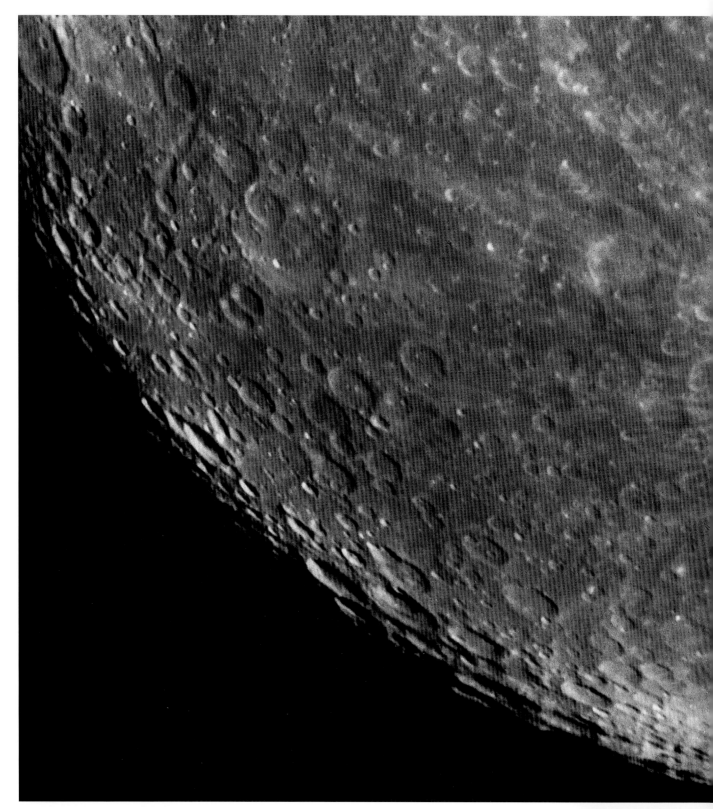

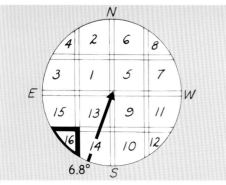

2326 UT 06.02.66. Moon's age 16.3 days. Diameter 64 cm. This was taken nearly $1\frac{1}{2}$ days after Full Moon, and so some of the limb detail has already disappeared, despite the favourable libration.

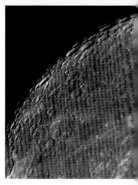

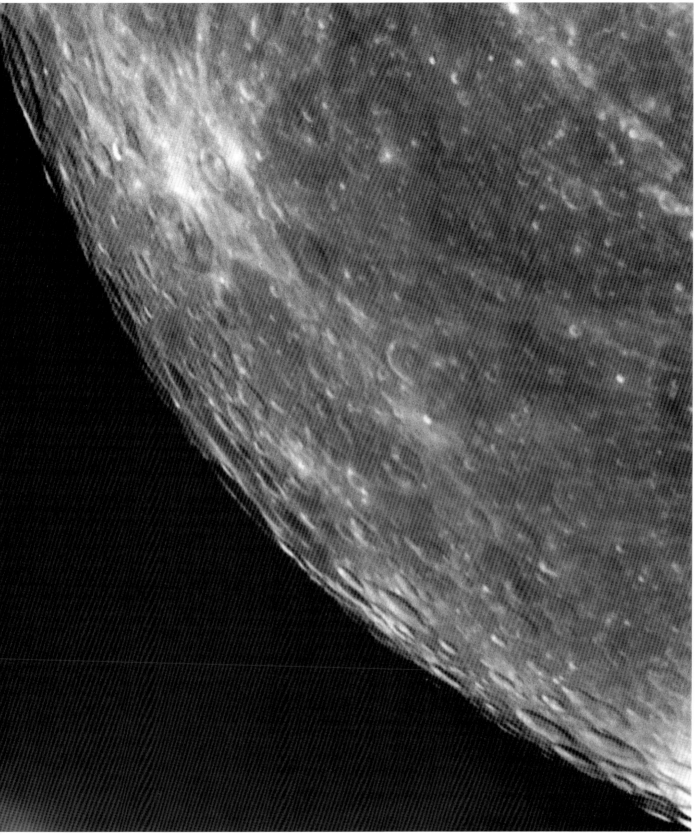

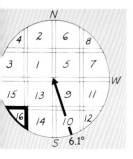

2259 UT 24.02.67. Moon's age 15.6 days. Diameter 64 cm. This was taken about 5 hours after Full Moon. The libration was not favourable, but even so the craters on the limb near Brisbane (d4) will not often be seen like this.

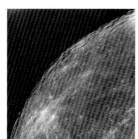

Plate 16d

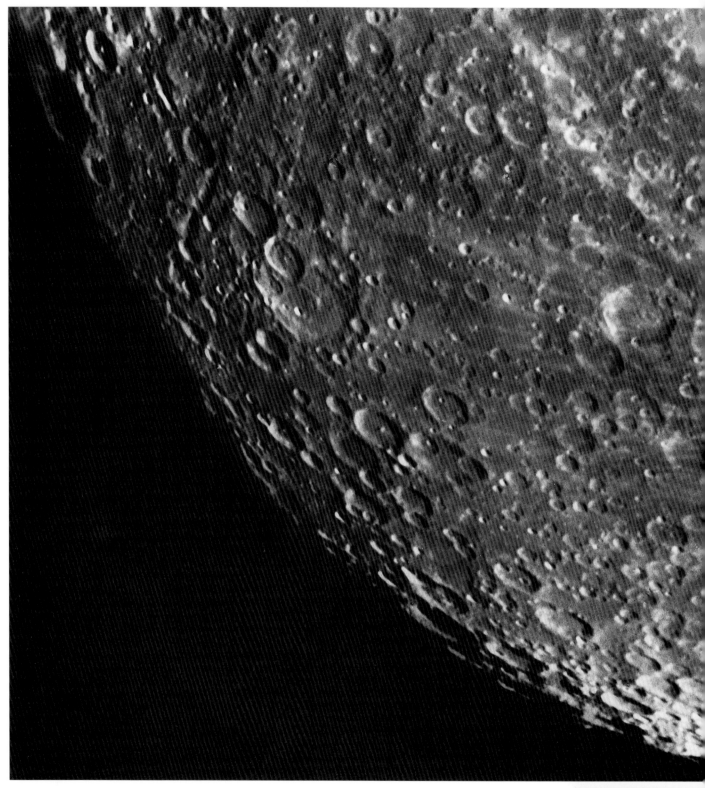

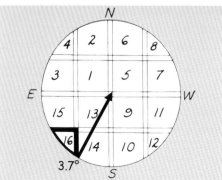

2334 UT 26.02.67. Moon's age 17.6 days. Diameter 64 cm. Compare this with Plate 16b, which shows almost exactly the same area.

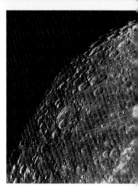

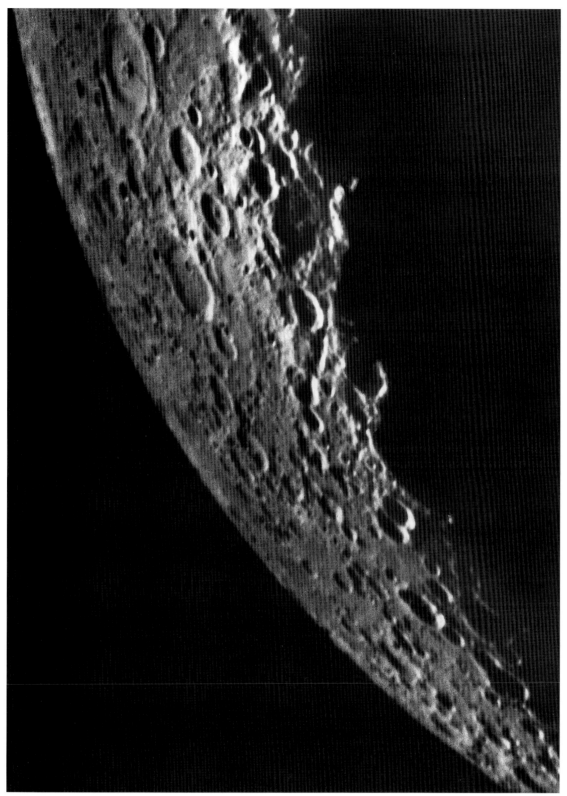

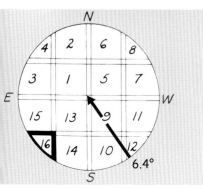

2036 UT 23.05.66. Moon's age 3.4 days. Diameter 64 cm. Compare this with Plate 16c. The librations are very similar.

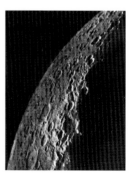

Plate 16f

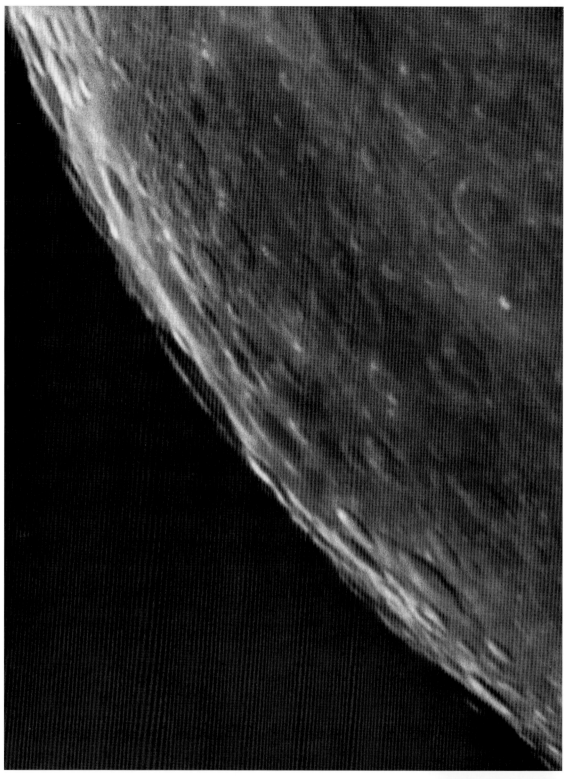

The Mare Australe Area (e3). 2259 UT 24.02.67. Moon's age 15.6 days. Diameter 99 cm approx. This is an enlargement of part of Plate 16c.

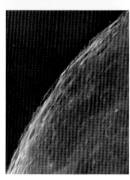

Plate 17

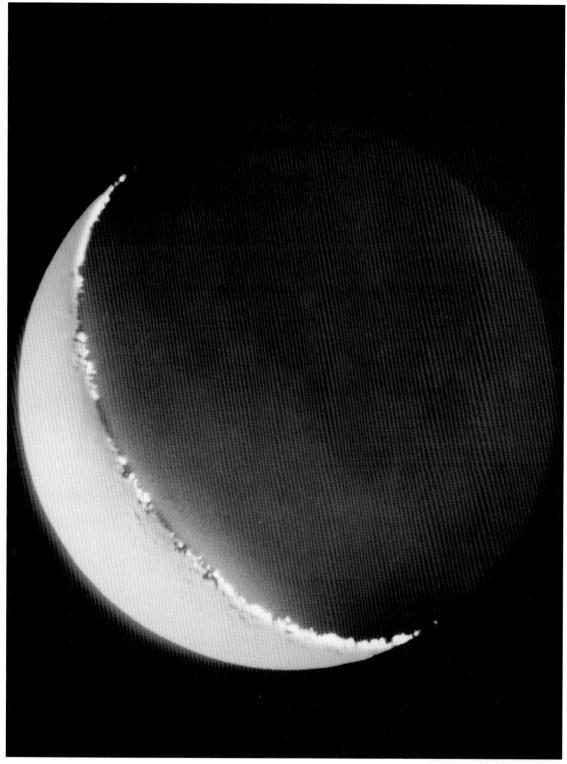

Earthshine—the Moon's surface illuminated by light reflected off the
Earth. 2032 UT 13.05.67. Moon's age 4.3 days. Compare this with the Full
Moon Key photograph, immediately preceding Map 1. The Earthshine
needed an exposure of 10 seconds, and the part of the Moon illuminated
by the Sun was therefore grossly over-exposed.

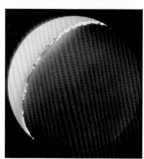

Table of Exposures

Date	UT	Focal ratio	Emulsion	Exposure sec.	Moon's age days	Plate Nos.
08.11.65	2132*	f/29	Kodak O.250 Plate	0.4	15.3	7e/2
08.01.66	2228	f/24	Kodak O.250 Plate	0.3†	17.0	3b, 15c
31.01.66	1947	f/24	Kodak O.250 Plate	0.3	10.1	5b
02.02.66	1944	f/24	Kodak O.250 Plate	0.25	12.1	7b, 8b
06.02.66	2326	f/24	Kodak O.250 Plate	0.2	16.3	14c, 16b
05.03.66	2259	f/41	Ilford Zenith Plate	0.1	13.5	10e/1
27.04.66	2152	f/24	Ilford G.30 Plate	0.5	6.9	15a
28.04.66	2046	f/30	Ilford G.30 Plate	0.8	7.9	1a, 13a
23.05.66	2033	f/24	Kodak O.250 Plate	0.7	3.4	15e/2
23.05.66	2034	f/24	Kodak O.250 Plate	0.7	3.4	3e/2, 4b
23.05.66	2036	f/24	Kodak O.250 Plate	0.7	3.4	16e
28.05.66	2122	f/30	Ilford G.30 Plate	0.8	8.4	14a
29.05.66	2103	f/30	Ilford G.30 Plate	0.8	9.4	2c
23.06.66	2109	f/24	Ilford G.30 Plate	1.0	5.0	16a
06.08.66	0211	f/30	Ilford G.30 Plate	0.7	18.9	2b
06.08.66	0215	f/30	Ilford G.30 Plate	0.7	18.9	1b, 13b
09.08.66	0315	f/30	Ilford G.30 Plate	1.1	23.8	5a,6a
09.08.66	0317	f/30	Ilford G.30 Plate	1.1	23.8	9a, 9f/2, 11b, 12c/1
06.10.66	0516	f/30	Ilford G.30 Plate	1.5	21.4	9d, 10d, 10e/2
28.10.66	2219	f/30	Ilford G.30 Plate	0.4	14.7	7e/1, 8c
28.10.66	2229	f/30	Ilford G.30 Plate	0.4	14.7	11c, 12d
29.10.66	2248	f/30	Ilford G.30 Plate	0.3	15.7	4d
04.11.66	0609	f/30	Ilford G.30 Plate	1.5	21.0	10a
22.11.66	1814	f/30	Ilford G.30 Plate	0.8	10.2	2a
23.11.66	2142	f/30	Ilford G.30 Plate	0.7	11.3	6c
28.11.66	2125	f/30	Ilford G.30 Plate	0.5	16.3	15b
28.11.66	2127	f/30	Ilford G.30 Plate	0.5	16.3	4c
01.12.66	2336	f/30	Ilford G.30 Plate	0.7	19.4	14b
07.12.66	0655	f/30	Ilford G.30 Plate	2.5‡	24.6	7c,8d
23.12.66	2232	f/30	Ilford G.30 Plate	0.6	11.8	6d
23.12.66	2236	f/30	Ilford G.30 Plate	0.6	11.8	9e
25.12.66	2036	f/30	Ilford G.30 Plate	0.4	13.8	7d, 11d, 11e/1, 12b
25.12.66	2040	f/30	Ilford G.30 Plate	0.4	13.8	6b, 8a
25.12.66	2135	f/30	Ilford G.30 Plate	0.4	13.8	5c, 9c
25.12.66	2137	f/30	Ilford G.30 Plate	0.4	13.8	1c
21.01.67	1758	f/30	Ilford G.30 Plate	0.5	11.0	9b, 10b
24.01.67	2310	f/30	Ilford G.30 Plate	0.4	14.2	7a, 11a
24.01.67	2314	f/30	Ilford G.30 Plate	0.4	14.2	13d
16.02.67	1757	f/30	Ilford G.30 Plate	1.0	7.3	3a, 4a
19.02.67	1801	f/30	Gevaert R.23 Plate	0.5	10.3	5d
23.02.67	2104	f/30	Ilford G.30 Plate	0.6†	14.5	10c, 12a, 12e/1
24.02.67	2259	f/30	Ilford G.30 Plate	0.6†	15.6	16c, 16f
24.02.67	2300	f/30	Ilford G.30 Plate	0.6†	15.6	15d, 15e/1
24.02.67	2353	f/30	Ilford G.30 Plate	0.6†	15.6	3d
26.02.67	2334	f/30	Ilford G.30 Plate	0.6†	17.6	13c, 16d
26.02.67	2336	f/30	Ilford G.30 Plate	0.6†	17.6	3c
18.03.67	1835	f/30	Ilford G.30 Plate	1.0	7.6	1d,2d
19.03.67	1925	f/60	Ilford FP3 35 mm	1.0	8.6	2e/2
19.03.67	1928	f/60	Ilford FP3 35 mm	1.0	8.6	9f/1
19.03.67	1959	f/30	Ilford G.30 Plate	0.8	8.6	13f, 14d
20.03.67	2029	f/60	Ilford FP3 35 mm	1.0	9.7	6e/1
20.03.67	2041	f/60	Ilford FP3 35 mm	1.0	9.7	5e/1
20.03.67	2046	f/60	Ilford FP3 35 mm	1.0	9.7	14e/2
20.03.67	2050	f/60	Ilford FP3 35 mm	0.5	9.7	13g/2
20.03.67	2056	f/60	Ilford FP3 35 mm	1.0	9.7	9g/1
21.03.67	1908	f/60	Ilford FP3 35 mm	1.0	10.6	5e/2
21.03.67	1923	f/60	Ilford FP3 35 mm	1.0	10.6	6e/2
21.03.67	1952	f/60	Ilford FP3 35 mm	1.0	10.6	9g/2
23.03.67	1938	f/60	Ilford FP3 35 mm	1.0	12.6	12e/2
23.03.67	1944	f/60	Ilford FP3 35 mm	1.0	12.6	8e/2

Date	UT	Focal ratio	Emulsion	Exposure sec.	Moon's age days	Plate Nos.
17.04.67	1950	f/52	Ilford Pan F 35 mm	2.0	7.9	13e/2
17.04.67	1951	f/52	Ilford Pan F 35 mm	2.0	7.9	13g/1
20.04.67	2106	f/52	Ilford Pan F 35 mm	2.0	11.0	11e/2
21.04.67	1952	f/52	Ilford Pan F 35 mm	0.5	11.9	8e/3b
21.04.67	1958	f/52	Ilford Pan F 35 mm	2.0	11.9	12c/2
21.04.67	2033	f/52	Ilford Pan F 35 mm	2.0	12.0	8e/1
23.04.67	2150	f/52	Ilford Pan F 35 mm	1/8	14.0	8e/3c
13.05.67	2002	f/52	Ilford Pan F 35 mm	3.0	4.2	3g/1
13.05.67	2032	f/7.25§	Ilford Pan F 35 mm	10.0	4.3	17
15.05.67	2000	f/52	Ilford Pan F 35 mm	3.0	6.2	3f/2
15.05.67	2008	f/52	Ilford Pan F 35 mm	2.0	6.2	3g/2
15.05.67	2015	f/52	Ilford Pan F 35 mm	3.0	6.2	3f/1
16.05.67	2008	f/52	Ilford Pan F 35 mm	2.0	7.2	13e/1
16.05.67	2012	f/52	Ilford Pan F 35 mm	3.0	7.2	2e/1
16.05.67	2016	f/52	Ilford Pan F 35 mm	3.0	7.2	1e/1
16.05.67	2017	f/52	Ilford Pan F 35 mm	3.0	7.2	1e/2
16.05.67	2021	f/52	Ilford Pan F 35 mm	2.0	7.3	14e/1
20.05.67	2110	f/52	Ilford Pan F 35 mm	0.5	11.3	8e/3a
20.08.67	2304	f/30	Ilford G.30 Plate	0.3	14.9	3e/1, 4e

Except where noted the instrument was a 30 cm Newtonian Reflector with a silvered mirror.

* The instrument here was a 15 cm Newtonian Reflector with a silvered mirror
‡ An Ilford α pale yellow filter was used
† This was taken through mist; hence the long exposure
§ This was taken at the Prime Focus

Emulsion speeds:		
	Ilford FP3 35 mm	ASA 120
	Ilford Pan F 35 mm	ASA 50
	Ilford Zenith Plate	ASA 50
	Ilford G.30 Plate	ASA 10
	Kodak O.250 Plate	ASA 25
	Gevaert R.23 Plate	ASA 12

Note that although these are the makers quoted speeds, they do not really apply to the fairly long exposures used here due to reciprocity failure. Despite some of these emulsions not now being available, the combination of speed, focal ratio and corresponding exposure are a good starting point for acceptable photographs using modern materials.

Index of Named Formations

This index contains formations listed in NASA Catalogue of Lunar Nomenclature by L. E Andersson and E. A. Whitaker (NASA Reference Publication 1097) (1982) with additions to 1994. A few older, unofficial, names are indicated in this index with an asterisk (*) while on the maps, they are shown enclosed between brackets.

Formations such as Mares, Montes, Promontoria etc are additionally indexed under their proper names. The groups of figures and letters against the formations indicate where they will be found on the various key maps. Thus the group 3 b4 against Alhazen indicates that it lies in square b4 on Map 3: Adams, 15 b1, 16 b7 may be found on Maps 15 and 16.

Feature	Map reference	Lat	Long	km
Abbot	3 c2	5.6°N	54.8°E	10
Abenezra	13 d3	21.0°S	11.9°E	42
Abulfeda	13 c5, 13 d5	13.8°S	13.9°E	65
Acosta	15 b6, 15 c6	5.6°S	60.1°E	13
Adams	15 d1, 16 b7	31.9°S	68.2°E	66
Aestatis, Lacus	11 g5	15°S	69°W	90
Aestuum, Sinus	5 c4	12°N	8°W	230
Agarum, Promontorium	3 b4	14°N	66°E	70
Agassiz, Promontorium	2 f3	42°N	2°E	20
Agatharchides	9 f4, 11 b3	19.8°S	30.9°W	49
Agricola, Montes	8 e2	29°N	54°W	160
Agrippa	1 e2, 13 d8	4.1°N	10.5°E	44
Airy	13 e4	18.1°S	5.7°E	37
Al-Bakri	1 c4	14.3°N	20.2°E	12
Al-Marrakushi	15 c5, 15 c6	10.4°S	55.8°E	8
Albategnius	13 e5, 13 f5	11.2°S	4.1°E	136
Alexander	2 d3	40.3°N	13.5°E	82
Alfraganus	13 b6	5.4°S	19.0°E	21
Alhazen	3 b4	15.9°N	71.8°E	33
Aliacensis	10 a8, 13 f1, 14 e8	30.6°S	5.2°E	80
Almanon	13 c4	16.8°S	15.2°E	49
Alpes, Montes	2 f4, 2 g5, 6 a5	46°N	1°W	250
Alpes, Vallis	2 f5	49°N	3°E	180
Alpetragius	9 a4, 13 g4	16.0°S	4.5°W	40
Alphonsus	9 a4, 13 g5	13.4°S	2.8°W	119
Alpine Valley (Vallis Alpes)	2 f5	49°N	3°E	180
Altai, Rupes	13 b2, 14 a8, 15 h2, 16 g8	24°S	23°E	480
Ameghino	3 b1	3.3°N	57.0°E	9
Ammonius	13 g6	8.5°S	0.8°W	9
Ampère, Mons	5 b6	19°N	4°W	30
Amundsen	14 d1	84.5°S	82.8°E	105
Anaxagoras	2 g7, 2 g8, 6 b8	73.4°N	10.1°W	51
Anaximander	6 e7	66.9°N	51.3°W	68
Anaximenes	6 d8	72.5°N	44.5°W	80
Andel	13 d5	10.4°S	12.4°E	35
Angström	5 h8, 6 h2, 8 c3	29.9°N	41.6°W	10
Anguis, Mare	3 b6, 4 b1	22°N	67°E	130
Ansgarius	15 a5	12.7°S	79.7°E	94
Anville	3 c2	1.9°N	49.5°E	11
Apenninus, Montes	1 g5, 5 a6	20°N	3°W	600
Apianus	13 e2, 14 e8	26.9°S	7.9°E	63
Bond, W	2 f7, 6 a7	65.3°N	3.7°E	158
Bonpland	9 c6	8.3°S	17.4°W	60
Boole	6 g7, 8 b8	63.7°N	87.4°W	63
Borda	15 e3	25.1°S	46.6°E	44
Borel	1 b5, 3 g6, 4 g1	22.3°N	26.4°E	5

Feature	Map reference	Lat	Long	km
Apollonius	3 b2	4.5°N	61.1°E	53
Arago	1 c2, 3 h3	6.2°N	21.4°E	26
Aratus	1 f6	23.6°N	4.5°E	11
Archerusia, Promontorium	1 c4, 3 h5	17°N	22°E	10
Archimedes	1 h7, 2 g1, 5 b8, 6 a2	29.7°N	4.0°W	83
Archimedes, Montes	1 h7, 5 b7, 6 b1	26°N	5°W	140
Archytas	2 f6	58.7°N	5.0°E	32
Argaeus, Mons	1 b5, 3 g6, 4 g1	19°N	29°E	50
Argelander	13 e4	16.5°S	5.8°E	34
Ariadaeus	1 d2	4.6°N	17.3°E	11
Ariadaeus, Rima	1 d2	7°N	13°E	220
Aristarchus	7 c8, 7 d8, 8 d2	23.7°N	47.4°W	40
Aristillus	1 g8, 2 f2, 5 a8, 6 a2	33.9°N	1.2°E	55
Aristoteles	2 d5	50.2°N	17.4°E	87
Arnold	2 c7, 2 d7	66.8°N	35.9°E	95
Artsimovich	5 h8, 6 g1, 7 b8, 8 c2	27.6°N	36.6°W	9
Aryabhata	3 e3, 3 f3	6.2°N	35.1°E	22
Arzachel	9 a3, 13 g4	18.2°S	1.9°W	97
Asada	3 c3	7.3°N	49.9°E	12
Asclepi	14 c4	55.1°S	25.4°E	43
Atlas	2 a4, 4 g6	46.7°N	44.4°E	87
Atwood	15 c7	5.8°S	57.7°E	29
Australe, Mare	16 e3	46°S	91°E	900
Autolycus	1 g7, 2 f1, 5 a8	30.7°N	1.5°E	39
Autumni, Lacus	11 h5	14°S	82°W	240
Auwers	1 d4	15.1°N	17.2°E	20
Auzout	3 b3	10.3°N	64.1°E	33
Azophi	13 d3	22.1°S	12.7°E	48
Babbage	6 g6, 8 b7	59.5°N	56.8°W	144
Baco	14 c4	51.0°S	19.1°E	70
Baillaud	2 d8	74.6°N	37.5°E	90
Bailly	10 g2, 12 d2	66.8°S	69.4°W	303
Baily	2 b5, 4 h6	49.7°N	30.4°E	27
Balboa	7 g7	19.1°N	83.2°W	70
Ball	10 c7, 10 d7, 14 h7	35.9°S	8.4°W	41
Banachiewicz	3 a2	5.2°N	80.1°E	92
Bancroft	2 h1, 5 b7, 6 b1	28.0°N	6.4°W	13
Banting	1 d6	26.6°N	16.4°E	5
Barkla	15 b5	10.7°S	67.2°E	43
Barocius	14 d5	44.9°S	16.8°E	82
Barrow	2 f7	71.3°N	7.7°E	93
Bartels	7 g7, 8 h1	24.5°N	89.8°W	55
Bayer	10 g4, 12 c4	51.6°S	35.0°W	47
Beaumont	13 a4, 15 g4	18.0°S	28.8°E	53
Beer	5 c7, 6 b1	27.1°N	9.1°W	10
Behaim	15 b4	16.5°S	79.4°E	55
Beketov	1 b4, 3 f5	16.3°N	29.2°E	8
Bellot	15 d5	12.4°S	48.2°E	17
Bernouilli	3 d8, 4 d4	35.0°N	60.7°E	47
Berosus	3 c8, 4 c4	33.5°N	69.9°E	74
Berzelius	4 e4	36.6°N	50.9°E	51
Bessarion	7 b6	14.9°N	37.3°W	10
Bessel	1 d5	21.8°N	17.9°E	16
Bettinus	10 g2, 10 h2, 12 c2	63.4°S	44.8°W	71
Bianchini	6 f5, 8 a6	48.7°N	34.3°W	38
Biela	16 f3	54.9°S	51.3°E	76
Bilharz	15 c6, 15 c7	5.8°S	56.3°E	43
Billy	11 e5	13.8°S	50.1°W	46
Biot	15 d3	22.6°S	51.1°E	13
Birmingham	2 g7, 6 b7	65.1°N	10.5°W	92
Birt	9 b3, 13 h3	22.4°S	8.5°W	17
Birt, Rima	9 b3	21°S	9°W	50
Black	15 a5	9.2°S	80.4°E	18
Blagg	1 g1, 13 f8	1.3°N	1.5°E	5
Blanc, Mons	2 f4	45°N	1°E	25
Blancanus	10 e2, 12 a2	63.6°S	21.5°W	105
Blanchinus	13 f2	25.4°S	2.5°E	61
Bobillier	1 d5	19.6°N	15.5°E	7
Bode	1 h3, 5 b3	6.7°N	2.4°W	19
Boguslawsky	14 c2	72.9°S	43.2°E	97
Bohnenberger	15 e5	16.2°S	40.0°E	33
Bohr	7 h5	12.8°N	86.4°W	71
Bombelli	3 c2	5.3°N	56.2°E	10
Bond, G	1 a7, 3 f8, 4 f3	32.4°N	36.2°E	20

Feature	Map reference	Lat	Long	km
Born	15 b6	6.0°S	66.8°E	15
Boscovich	1 e3	9.8°N	11.1°E	46
Bouguer	6 f5, 8 a6	52.3°N	35.8°W	23
Boussingault	14 b2, 16 h1	70.4°S	54.7°E	131
Bouvard, Vallis	12 h7	39°S	83°W	280
Bowen	1 e5	17.6°N	9.1°E	9
Bradley, Mons	1 g6	22°N	1°E	30
Brayley	5 h6, 7 b7, 8 c1	20.9°N	36.9°W	15
Breislak	14 c5	48.2°S	18.3°E	50
Brenner	16 f6	39.0°S	39.3°E	97
Brewster	1 a5, 3 f6, 4 f1, 4 f2	23.3°N	34.7°E	11
Brianchon	6 e8	74.8°N	86.5°W	145
Briggs	7 f8, 8 g2	26.5°N	69.1°W	37
Brisbane	16 d4	49.1°S	68.5°E	45
Brown	10 e5, 12 a4	46.4°S	17.9°W	34
Bruce	1 g1, 13 f8	1.1°N	0.4°E	7
Buch	14 c6	38.8°S	17.7°E	54
Bullialdus	9 e3, 9e4	20.7°S	22.2°W	61
Bunsen	7 f5	41.4°N	85.3°W	52
Burckhardt	3 d8, 4 d3	31.1°N	56.5°E	57
Bürg	2 b4, 4 h5	45.0°N	28.2°E	40
Burnham	13 e5	13.9°S	7.3°E	25
Büsching	14 c7	38.0°S	20.0°E	52
Byrgius	11 g3	24.7°S	65.3°W	87
C. Herschel	6 f2, 6 f3, 8 a4	34.5°N	31.2°W	13
C. Mayer	2 e6	63.2°N	17.3°E	38
Cabeus	10 d1	84.9°S	35.5°W	98
Cajal	1 a3, 3 f4	12.6°N	31.1°E	9
Calippus	2 e3	38.9°N	10.7°E	33
Cameron	3 d3	6.2°N	45.9°E	11
Campanus	9 f2, 10 g8, 11 a1, 12 b8	28.0°S	27.8°W	48
Capella	15 f7	7.6°S	34.9°E	49
Capuanus	9 f1, 10 g7, 12 b7	34.1°S	26.7°W	60
Cardanus	7 g5	13.2°N	72.4°W	50
Carlini	6 e2	33.7°N	24.1°W	11
Carmichael	3 e5, 4 e1	19.6°N	40.4°E	20
Carpatus, Montes	5 f5	15°N	25°W	400
Carpenter	6 e7	69.4°N	50.9°W	60
Carrel	1 b3, 3 g4	10.7°N	26.7°E	16
Carrington	4 e5	44.0°N	62.1°E	30
Cartan	3 b2	4.2°N	59.3°E	16
Casatus	10 e1, 10 e2, 12 a1	72.6°S	30.5°W	111
Cassini	2 f3	40.2°N	4.6°E	57
Catharina	13 b4, 15 h4	18.0°S	23.6°E	100
Caucasus, Montes	1 f8, 2 e2	39°W	9°E	520
Cauchy	3 e3	9.6°N	38.6°E	12
Cavalerius	7 g4	5.1°N	66.8°W	58
Cavendish	11 e3	24.5°S	53.7°W	56
Caventou	5 f8, 6 f2, 8 a3	29.8°N	29.4°W	3
Cayley	1 d2, 13 c8	4.0°N	15.1°E	14
Celsius	14 c7	34.1°S	20.1°E	36
Censorinus	3 f1, 15 f8	0.4°S	32.7°E	4
Cepheus	4 f5	40.8°N	45.8°E	40
Chacornac	1 b7, 2 a1, 3 f8, 4 g3	29.8°N	31.7°E	51
Challis	2 f8	79.5°N	9.2°E	56
Chevallier	4 f5	44.9°N	51.2°E	52
Chladni	1 g2	4.0°N	1.1°E	14
Cichus	9 e1, 10 f7, 12 a7	33.3°S	21.1°W	41
Clairaut	14 d5	47.7°S	13.9°E	75
Clausius	9 h1, 12 e7	36.9°S	43.8°W	25
Clavius	10 d3, 12 a2, 14 g3, 14h3	58.4°S	14.4°W	225
Cleomedes	3 c7, 3 d7, 4 d2	27.7°N	55.5°E	126
Cleostratus	6 g7, 8 b7, 8 c7	60.4°N	77.0°W	63
Clerke	1 a5, 3 f6, 4 g1	21.7°N	29.8°E	7
Cognitum, Mare	9 e6	10°S	23°W	340
Colombo	15 d5	15.1°S	45.8°E	76
Condon	3 b1, 15 b8	1.9°N	60.4°E	35
Condorcet	3 b3	12.1°N	69.6°E	74
Conon	1 g6	21.6°N	2.0°E	22
Cook	15 d4	17.5°S	48.9°E	47
Copernicus	5 e4	9.7°N	20.0°W	93
Cordillera, Montes	11 h5	20°S	80°W	1,500
Cremona	6 f7	67.5°N	90.6°W	85
Crile	3 d4	14.2°N	46.0°E	9

Feature	Map reference	Lat	Long	km
Crisium, Mare	3 c5, 4 c1	17°N	59°E	570
Crozier	15 d5	13.5°S	50.8°E	22
Crüger	11 g5	16.7°S	66.8°W	46
Curtius	10 b2, 14 e2	67.2°S	4.4°E	95
Cusanus	2 c8	72.0°N	70.8°E	63
Cuvier	10 a4, 14 e4	50.3°S	9.9°E	75
Cyrillus	13 a5, 13 b5, 15 h5	13.2°S	24.0°E	98
Cysatus	10 c2, 14 f2	66.2°S	6.1°W	49
d'Arrest	1 d1, 13 c8	2.3°N	14.7°E	30
da Vinci	3 d3	9.1°N	45.0°E	38
Daguerre	15 f6,	11.9°S	33.6°E	46
Dalton	7 g6	17.1°N	84.3°W	61
Daly	3 b2	5.7°N	59.6°E	17
Damoiseau	7 f2, 11 f7	4.8°S	61.1°W	37
Daniell	1 b8, 2 a2, 4 g4	35.3°N	31.1°E	29
Darney	9 e5	14.5°S	23.5°W	15
Darwin	11 g4	19.8°S	69.1°W	130
Daubrée	1 d4	15.7°N	14.7°E	14
Davy	9 b5, 13 h5	11.8°S	8.1°W	35
Dawes	1 b4, 3 g5	17.2°N	26.4°E	18
de Gasparis	11 e2	25.9°S	50.7°W	30
de la Rue	2 a6, 4 h7	59.1°N	53.0°E	136
de Morgan	1 d1, 1 d2, 13 c8	3.3°N	14.9°E	10
de Sitter	2 e8	80.1°N	39.6°E	65
de Vico	11 f4	19.7°S	60.2°W	20
Debes	3 d7, 4 d3	29.5°N	51.7°E	31
Dechen	8 d5	46.1°N	68.2°W	12
Delambre	13 c7	1.9°S	17.5°E	52
Delaunay	13 f3	22.2°S	2.5°E	46
Delisle	5 g8, 6 g2, 8 b3	29.9°N	34.6°W	25
Delisle, Mons	5 g8, 6 g2, 8 b3	29°N	36°W	30
Delmotte	3 c7, 4 c2	27.1°N	60.2°E	33
Deluc	10 c4, 14 f3	55.0°S	2.8°W	47
Dembowski	1 f2, 13 e8	2.9°N	7.2°E	26
Democritus	2 c6	62.3°N	35.0°E	39
Demonax	14 c1	78.2°S	59.0°E	114
Desargues	6 f8	70.2°N	73.3°W	85
Descartes	13 c5	11.7°S	15.7°E	48
Deseilligny	1 c5, 3 h6, 4 h1	21.1°N	20.6°E	6
Deslandres	9 b1, 10 c7, 13 h1, 14 g7	32.5°S	5.2°W	234
Deville, Promontorium	2 f4	43°N	1°E	20
Dionysius	1 d1, 13 c8	2.8°N	17.3°E	18
Diophantus	5 g8, 6 g1, 7 a8, 8 b2	27.6°N	34.3°W	18
Dollond	13 c5	10.4°S	14.4°E	11
Donati	13 e3	20.7°S	5.2°E	36
Doppelmayer	9 h3, 11 c2, 11 d2, 12 d8	28.5°S	41.4°W	64
Dorsa Lister	3 g6, 4 g1	19°N	22°E	290
Dorsa Smirnov	3 g7, 4 g2	25°N	25°E	130
Dove	14 b5, 16 g4, 16 g5	46.7°S	31.5°E	30
Draper	5 e5, 5 e6	17.6°N	21.7°W	8
Drebbel	12 e6	40.9°S	49.0°W	30
Drygalski	10 e1	79.7°S	86.8°W	163
Dubyago	3 b2	4.4°N	70.0°E	51
Dunthorne	9 g2, 10 h8, 11 b1, 12 c8	30.1°S	31.6°W	16
Eddington	7 f7, 8 g1	21.5°N	71.8°W	125
Egede	2 e4	48.7°N	10.6°E	37
Eichstadt	11 h3	22.6°S	78.3°W	49
Eimmart	3 b6, 4 c2	24.0°N	64.8°E	46
Einstein	7 h6	16.6°N	88.5°W	170
Elger	9 f1, 10 g7, 12 c7	35.3°S	29.8°W	21
Encke	7 c4, 11 b8	4.6°N	36.6°W	28
Endymion	4 f6, 4 g7	53.6°N	56.5°E	125
Epidemiarum, Palus	9 f2, 10 g7, 12 b7	32°S	27°W	300
Epigenes	2 g7, 6 b7	67.5°N	4.6°W	55
Epimenides	10 g6, 12 b6	40.9°S	30.2°W	27
Eppinger *	9 e6	9.4°S	25.7°W	6
Eratosthenes	5 c5	14.5°N	11.3°W	58
Esclangon	3 e6, 4 e1	21.5°N	42.1°E	16
Euclides	7 a1, 9 f7, 11 a6	7.4°S	29.5°W	11
Euctemon	2 e8	76.4°N	31.3°E	62
Eudoxus	2 d4	44.3°N	16.3°E	67
Euler	5 g7, 8 a2	23.3°N	29.2°W	28
Fabbroni	1 b4	18.7°N	29.2°E	11
Fabricius	16 f5	42.9°S	42.0°E	78

Feature	Map reference	Lat	Long	km
Fahrenheit	3 b4	13.1°N	61.7°E	6
Faraday	10 a6, 14 e6	42.4°S	8.7°E	70
Fauth	5 e3	6.3°N	20.1°W	12
Faye	13 f3	21.4°S	3.9°E	37
Fecunditatis, Mare	3 c1, 15 c7	8°S	51°E	840
Fermat	13 c3	22.6°S	19.8°E	39
Fernelius	10 a6, 14 e6	38.1°S	4.9°E	65
Feuillée	2 h1, 5 c7, 6 b1	27.4°N	9.4°W	9
Firmicus	3 b3	7.3°N	63.4°E	56
Flammarion	5 b1, 13 g7	3.4°S	3.7°W	75
Flamsteed	7 d2, 9 h8, 11 d7	4.5°S	44.3°W	21
Fontana	11 f5	16.1°S	56.6°W	31
Fontenelle	2 h6, 6 c7	63.4°N	18.9°W	38
Foucault	6 f5, 8 a6	50.4°N	39.7°W	23
Fourier	11 e1, 12 f8	30.3°S	53.0°W	52
Fra Mauro	5 e1, 9 c6	6.0°S	17.0°W	95
Fracastorius	15 g4	21.2°S	33.0°E	124
Franck	1 a5, 3 f6, 4 f1	22.6°N	35.5°E	12
Franklin	4 e4	38.8°N	47.7°E	56
Franz	3 e5	16.6°N	40.2°E	26
Fraunhofer	16 d6	39.5°S	59.1°E	57
Fredholm	3 d5, 4 d1	18.4°N	46.5°E	15
Fresnel, Promontorium	1 f7, 2 e1	29°N	5°E	20
Frigoris, Mare	2 e5, 6 c6	55°N	5°E	1,350
Furnerius	15 d1, 16 c7	36.3°S	60.4°E	125
G. Bond	1 a7, 3 f8, 4 f3	32.4°N	36.2°E	20
Galen	1 f6	21.9°N	5.0°E	10
Galilaei	7 f5	10.5°N	62.7°W	16
Galle	2 d6	55.9°N	22.3°E	21
Galvani	8 e6	49.6°N	84.6°W	80
Gambart	5 d2, 9 c8	1.0°N	15.2°W	25
Gardner	1 a4	17.7°N	33.8°E	18
Gärtner	2 c6	59.1°N	34.6°E	102
Gassendi	9 h5, 11 c4	17.5°S	39.9°W	110
Gaudibert	15 e6	10.9°S	37.8°E	34
Gauricus	9 c1, 10 d7	33.8°S	12.6°W	79
Gauss	3 c8, 4 c4	35.9°N	79.1°E	177
Gay-Lussac	5 e5	13.9°N	20.8°W	26
Gay-Lussac, Rima	5 e5	13°N	22°W	40
Geber	13 d3	19.4°S	13.9°E	45
Geminus	3 d8, 4 d4	34.5°N	56.7°E	86
Gemma Frisius	14 d7	34.2°S	13.3°E	88
Gerard	8 e5	44.5°N	80.0°W	90
Gibbs	15 b3	18.4°S	84.3°E	77
Gilbert	15 a7	3.2°S	76.0°E	107
Gill	16 f2	63.9°S	75.9°E	66
Gioja	2 f8	83.3°N	2.0°E	42
Glaisher	3 c4, 3 d4	13.2°N	49.5°E	16
Glushko	7 h4	8.1°N	77.6°W	43
Goclenius	15 d6	10.0°S	45.0°E	72
Godin	1 e1, 13 d8	1.8°N	10.2°E	35
Goldschmidt	2 f7, 2 g7, 6 b8	73.0°N	2.9°W	120
Golgi	7 e8, 8 f2	27.8°N	60.0°W	5
Goodacre	13 d1, 14 d8	32.7°S	14.1°E	46
Gould	9 d4	19.2°S	17.2°W	34
Greaves	3 c4	13.2°N	52.7°E	14
Grimaldi	7 g1, 7 g2, 11 g7	5.2°S	68.6°W	410
Grove	1 b8, 2 a3, 4 g5	40.3°N	32.9°E	28
Gruemberger	10 c2, 14 g2	66.9°S	10.0°W	94
Gruithuisen	6 g2, 8 c3	32.9°N	39.7°W	16
Gruithuisen Delta	6 g3, 8 b4	36°N	39°W	20
Gruithuisen Gamma	6 g3, 8 b5	37°N	41°W	20
Guericke	9 c5	11.5°S	14.1°W	64
Gutenberg	15 e6	8.6°S	41.2°E	74
Gyldén	13 f7	5.3°S	0.3°E	47
Hadley, Mons	1 f6	27°N	5°E	25
Haemus, Montes	1 c4, 3 h5	17°N	13°E	400
Hagecius	14 a3, 16 g3	59.8°S	46.6°E	76
Hahn	3 c7, 4 c3	31.3°N	73.6°E	84
Haidinger	10 f6, 12 b6	39.2°S	25.0°W	22
Hainzel	10 h6, 12 c6	41.3°S	33.5°W	70
Hall	1 a7, 3 f8, 4 f3, 4 f4	33.7°N	37.0°E	35
Halley	13 e6	8.0°S	5.7°E	36
Hanno	16 e3	56.3°S	71.2°E	56

Feature	Map reference	Lat	Long	km
Hansen	3 b4	14.0°N	72.5°E	40
Hansteen	11 e5	11.5°S	52.0°W	45
Hansteen, Mons	11 e5	12°S	50°W	30
Harbinger, Montes	6 h1, 7 b8, 8 c2	27°N	41°W	90
Harding	8 e5	43.5°N	71.7°W	23
Hargreaves	15 b7	2.2°S	64.0°E	16
Harpalus	6 f5, 8 a6	52.6°N	43.4°W	39
Hartwig	11 h7	6.1°S	80.5°W	79
Hase	15 d1, 16 c8	29.4°S	62.5°E	83
Hausen	10 h2, 12 e2	65.5°S	88.4°W	167
Hecataeus	15 b3	21.8°S	79.6°E	127
Hedin	15 b3	2.9°N	76.5°W	143
Heinrich	5 d7, 6 d1	24.8°N	15.3°W	7
Heinsius	10 e6, 12 a5	39.5°S	17.7°W	64
Heis	5 g8, 6 f2, 8 a3, 8 b3	32.4°N	31.9°W	14
Helicon	6 e4	40.4°N	23.1°W	25
Hell	9 b1, 10 c7, 13 h1, 14 h7	32.4°S	7.8°W	33
Helmholtz	14 b2, 16 g1	68.1°S	64.1°E	95
Henry	11 f3	24.0°S	56.8°W	41
Henry Frères	11 f3	23.5°S	58.9°W	42
Heraclides, Promontorium	6 f4, 8 a5	41°N	33°W	50
Heraclitus	10 a4, 14 e4	49.2°S	6.2°E	90
Hercules	2 a4, 4 g6	46.7°N	39.1°E	69
Herigonius	9 g6, 11 b5	13.3°S	33.9°W	15
Hermann	7 f2, 11 e8	0.9°S	57.3°W	16
Herodotus	7 d8, 8 e1	23.2°N	49.7°W	35
Herodotus, Mons	7 d8, 8 e2	27°N	53°W	5
Herschel	13 g6, 13 g7	5.7°S	2.1°W	41
Herschel, C	6 f2, 6 f3, 8 a4	34.5°N	31.2°W	13
Herschel, J	6 e7	62.1°N	41.2°W	156
Hesiodus	9 d2, 10 e8	29.4°S	16.3°W	43
Hesiodus, Rima	9 e2	30°S	21°W	300
Hevelius	7 g3, 11 f8	2.2°N	67.3°W	106
Hiemalis, Lacus	7 h2, 11 h8	15°N	14°E	50
Hill	3 e6, 4 e1	20.9°N	40.8°E	16
Hind	13 e6	7.9°S	7.4°E	29
Hippalus	9 f3, 11 b2	24.8°S	30.2°W	58
Hipparchus	13 e7	5.5°S	4.8°E	151
Holden	15 c4	19.1°S	62.5°E	47
Hommel	14 b4, 16 h3	54.6°S	33.0°E	125
Hooke	4 e5	41.2°N	54.9°E	37
Horrebow	6 e6	58.7°N	40.8°W	24
Horrocks	13 e7	4.0°S	5.9°E	31
Hortensius	5 g3, 7 a4	6.5°N	28.0°W	1
Huggins	10 b6, 14 f6	41.1°S	1.4°W	65
Humason	8 e3	30.7°N	56.6°W	4
Humboldt	15 c1, 16 b8	27.2°S	80.9°E	207
Humboldtianum, Mare	4 f7	57°N	80°E	160
Humorum, Mare	9 g4, 11 c3	24°S	39°W	400
Huxley	5 c6	20.2°N	4.5°W	4
Huygens, Mons	5 b6	20°N	3°W	40
Hyginus	1 f3	7.8°N	6.3°E	9
Hyginus, Rima	1 f3	8°N	6°E	220
Hypatia	3 h1, 13 b7, 15 h7	4.3°S	22.6°E	40
Ibn-Battuta	15 d6	6.9°S	50.4°E	12
Ibn-Rushd	13 b5, 15 h6	11.7°S	21.7°E	33
Ideler	14 c5	49.2°S	22.3°E	39
Imbrium, Mare	2 h3, 5 e7, 6 d3, 8 a2, 8 a3	35°N	15°W	1,300
Inghirami	12 f5	47.5°S	68.8°W	91
Insularum, Mare	9 d7	7°N	22°W	900
Iridum, Sinus	6 e4, 8 a5	45°N	32°W	200
Isidorus	15 f6	8.0°S	33.5°E	42
J. Herschel	6 e7	62.1°N	41.2°W	156
Jacobi	10 a3, 14 e3	56.7°S	11.4°E	68
Jansen	1 b3, 3 g4	13.5°N	28.7°E	24
Janssen	14 a5, 16 f5	44.9°S	41.5°E	190
Jenkins	3 a1, 15 a7	0.3°N	78.1°E	38
Joy	1 f6	25.0°N	6.6°E	6
Julius Caesar	1 d3	9.0°N	15.4°E	91
Jura, Montes	6 e5, 8 a6	47°N	37°W	380
Kaiser	10 a7, 14 e7	36.5°S	6.5°E	52
Kane	2 d6, 2 d7	63.1°N	26.1°E	55
Kant	13 b5	10.6°S	20.1°E	33
Kapteyn	15 b5	10.8°S	70.6°E	49

Feature	Map reference	Lat	Long	km
Kästner	15 a6	7.0°S	79.1°E	105
Keldysh	2 a5, 4 g6	51.2°N	43.6°E	33
Kelvin, Promontorium	9 g3, 11 b2	27°S	33°W	50
Kepler	7 c5	8.1°N	38.0°W	32
Kies	9 e2, 10 f8, 12 a8	26.3°S	22.5°W	46
Kinau	10 a3, 14 d3	60.8°S	15.1°E	42
Kirch	2 g3, 6 b3	39.2°N	5.6°W	12
Kircher	10 f2, 12 c2	67.1°S	45.3°W	73
Kirchhoff	3 e8, 4 f3	30.3°N	38.8°E	25
Klaproth	10 e2, 12 a1	69.7°S	26.0°W	119
Klein	13 f5	12.0°S	2.6°E	44
König	9 e3, 11 a2	24.1°S	24.6°W	23
Krafft	7 g6	16.6°N	72.6°W	51
Krieger	8 d3	29.0°N	45.6°W	22
Krogh	3 b3	9.4°N	65.7°E	20
Krusenstern	13 e2	26.2°S	5.9°E	47
Kuiper	9 d6, 9 e6	9.8°S	22.7°W	7
Kundt	9 b5	11.5°S	11.5°W	11
Kunowsky	5 h3, 7 b3, 7 b4, 11 a8	3.2°N	32.5°W	18
la Caille	13 f3	23.8°S	1.1°E	68
la Condamine	6 e6	53.4°N	28.2°W	37
la Hire, Mons	5 f8, 6 e1	28°N	25°W	25
la Pérouse	15 b5	10.7°S	76.3°E	78
Lacroix	12f7	37.9°S	59.0°W	38
Lacus Aestatis	11 g5	15°S	69°W	90
Lacus Autumni	11 h5	14°S	82°W	240
Lacus Hiemalis	7 h2, 11 h8	15°N	14°E	50
Lacus Mortis	2 b4, 4 h5	45°N	27°E	150
Lacus Somniorum	1 b8, 2 a3, 4 g4	36°N	31°E	230
Lacus Veris	11 h5	13°S	87°W	540
Lade	1 e1, 13 d7	1.3°S	10.1°E	56
Lagalla	10 f5, 12 b5	44.6°S	22.5°W	85
Lagrange	11 g1, 12 g8	33.2°S	72.0°W	160
Lalande	5 c1, 9 a6, 13 h7	4.4°S	8.6°W	24
Lamarck	11 g3	22.9°S	69.8°W	115
Lambert	5 e7, 6 e1	25.8°N	21.0°W	30
Lamé	15 b4, 15 c5	14.7°S	64.5°E	84
Laméch	2 d3	42.7°N	13.1°E	13
Lamont	1 b2, 1 c2, 3 g3, 3 h3	5.0°N	23.2°E	175
Landsteiner	5 d8, 6 c2	31.3°N	14.8°W	6
Langley	8 d6	51.1°N	86.3°W	60
Langrenus	15 c6	8.9°S	60.9°E	132
Lansberg	5 g2, 7 a3, 9 e8	0.3°S	26.6°W	39
Laplace, Promontorium	6 e5	47°N	26°W	50
Lassell	9 b4, 13 h4	15.5°N	7.9°W	23
Lavinium, Promontorium *	3 d4	15°N	49°E	2
Lavoisier	8 f4	38.2°N	81.2°W	70
Lawrence	3 d3	7.4°N	43.2°E	24
le Gentil	10 f1, 12 c1	74.4°S	76.5°W	113
le Monnier	1 b6, 3 f7, 4 g2	26.6°N	30.6°E	61
le Verrier	6 d4	40.3°N	20.6°W	20
Leakey	3 e1, 15 e7	3.2°S	37.4°E	13
Lee	9 h2, 11 c1, 12 d8	30.7°S	40.7°W	41
Legendre	15 c1, 16 b8	28.9°S	70.2°E	79
Lehmann	12 f6	40.0°S	56.0°W	53
Lepaute	9 g2, 10 h7, 12 c7	33.3°S	33.6°W	16
Letronne	7 d1, 9 h6, 9 h7, 11 c5, 11 d5	10.6°S	42.4°W	119
Lexell	10 c7, 13 g1, 14 g7	35.8°S	4.2°W	63
Licetus	10 a5, 14 e5	47.1°S	6.7°E	75
Lichtenberg	8 f3	31.8°N	67.7°W	20
Lick	3 c4	12.4°N	52.7°E	31
Liebig	11 e3	24.3°S	48.2°W	37
Liebig, Rupes	11 d3	25°S	46°W	180
Lilius	10 b4, 14 e4	54.5°S	6.2°E	61
Lindbergh	15 c7	5.4°S	52.9°E	13
Lindenau	13 b1, 14 b8, 16 h7	32.3°S	24.9°E	53
Lindsay	13 d6	7.0°S	13.0°E	32
Linné	1 e7, 2 d1	27.7°N	11.8°E	2
Lippershey	9 c2	25.9°S	10.3°W	7
Lister, Dorsa	3 g6	19°N	22°E	290
Littrow	1 a5, 3 f6, 4 f1	21.5°N	31.4°E	31
Lockyer	14 a5, 16 g5	46.2°S	36.7°E	34
Loewy	9 g4, 11 b3	22.7°S	32.8°W	24
Lohrmann	7 g2, 11 g8	0.5°S	67.2°W	31

Feature	Map reference	Lat	Long	km
Lohse	15 c5	13.7°S	60.2°E	42
Longomontanus	10 e4, 12 a4	49.5°S	21.7°W	145
Louville	6 g4, 8 b5	44.0°N	46.0°W	36
Lubbock	15 e7	3.9°S	41.8°E	14
Lubiniezky	9 e4	17.8°S	23.8°W	44
Lunicus, Sinus	1 g8, 2 f1, 2 f2	32°N	1°W	100
Luther	1 c7, 2 b2, 3 h8, 4 h3	33.2°N	24.1°E	10
Lyell	3 e4	13.6°N	40.6°E	32
Lyot	16 d4	50.2°S	84.1°E	141
Maclaurin	3 b1, 15 b7	1.9°S	68.0°E	50
Maclear	1 c3, 3 h4	10.5°N	20.1°E	20
Macmillan	5 c7, 6 b1	24.2°N	7.8°W	7
Macrobius	3 d6, 4 d1	21.3°N	46.0°E	64
Mädler	15 g6	11.0°S	29.8°E	28
Maestlin	7 c4	4.9°N	40.6°W	7
Magelhaens	15 e5	11.9°S	44.1°E	41
Maginus	10 c4, 14 g4	50.0°S	6.2°W	163
Main	2 f8	80.8°N	10.1°E	46
Mairan	6 g4, 8 c5	41.6°N	43.4°W	40
Malapert	10 c1, 14 e1, 14 f1	84.9°S	12.9°E	69
Mallet	16 e5	45.4°S	54.2°E	58
Manilius	1 e4, 1 f4	14.5°N	9.1°E	39
Manners	1 c2, 3 h3, 13 b8	4.6°N	20.0°E	15
Manzinus	14 d2	67.7°S	26.8°E	98
Maraldi	1 a4, 3 f6, 4 f1	19.4°N	34.9°E	40
Maraldi, Mons	3 f6, 4 f1	20°N	35°E	15
Marco Polo	1 h4, 5 a5	15.4°N	2.0°W	28
Mare Anguis	3 b6, 4 b1	22°N	67°E	130
Mare Australe	16 e3	46°S	91°E	900
Mare Cognitum	9 e6	10°S	23°W	340
Mare Crisium	3 c5, 4 c1	17°N	59°E	570
Mare Fecunditatis	3 c1, 15 c7	8°S	51°E	840
Mare Frigoris	2 e5, 6 c6	55°N	5°E	1,350
Mare Humboldtianum	4 f7	57°N	80°E	160
Mare Humorum	9 g4, 11 c3	24°S	39°W	400
Mare Imbrium	2 h3, 5 e7, 6 d3, 8 a2, 8 a3	35°N	15°W	1,300
Mare Insularum	9 d7	7°N	22°W	900
Mare Marginus	3 a4, 4 a1	12°N	88°E	360
Mare Nectaris	15 f5	15°S	35°E	350
Mare Nubium	9 d3	20°S	15°W	750
Mare Serenitatis	1 c6, 1 d6, 2 c1, 3 h7, 4 g2, 4 h2	27°N	19°E	650
Mare Smithii	15 a7	2°S	87°E	360
Mare Spumens	3 b1, 15 b8	1°N	65°E	130
Mare Tranquillitatis	1 b2, 3 f3, 13 a8, 15 g8	9°N	31°E	800
Mare Undarum	3 b2	7°N	69°E	220
Mare Vaporum	1 f4	13°N	3°E	230
Marginus, Mare	3 a4, 4 a1	12°N	88°E	360
Marinus	16 c6	39.4°S	76.5°E	58
Marius	7 e5	11.9°N	50.8°W	41
Markov	6 h6, 8 c6	53.4°N	62.7°W	40
Marth	9 f2, 10 g7, 11 b1, 12 b7	31.1°S	29.3°W	7
Maskelyne	1 a1, 3 f2	2.2°N	30.1°E	24
Mason	2 b4, 4 g5	42.6°N	30.5°E	34
Muupertuis	6 e5	49.6°N	27.3°W	46
Maurolycus	14 d6	41.8°S	14.0°E	114
Maury	1 a8, 4 f4	37.1°N	39.6°E	18
Mayer, C	2 e6	63.2°N	17.3°E	38
Mayer, T	5 g5, 7 a6	15.6°N	29.1°W	33
McClure	15 d5	15.3°S	50.3°E	24
McDonald	5 e8, 6 d2	30.4°N	20.9°W	8
Medii, Sinus	1 g1, 13 f8	3°N	3°E	350
Mee	10 g5, 10 h5, 12 c5	43.7°S	35.0°W	132
Menelaus	1 d4	16.3°N	16.0°E	27
Mercator	9 f2, 10 g8, 11 a1, 12 b8	29.3°S	26.1°W	47
Mercurius	4 e6	46.6°N	66.2°E	68
Mersenius	11 e3	21.5°S	49.2°W	84
Messala	4 d4	39.2°N	59.9°E	124
Messier	3 d1, 15 d8	1.9°S	47.6°E	11
Metius	16 e6, 16 f6	40.3°S	43.3°E	88
Meton	2 e7, 2 e8	73.8°N	19.2°E	122
Milichius	5 g4, 7 a5	10.0°N	30.2°W	12
Miller	10 b6, 14 f6	39.3°S	0.8°E	61
Mitchell	2 d5	49.7°N	20.2°E	30
Moigno	2 d7	66.4°N	28.9°E	37

Feature	Map reference	Lat	Long	km
Moltke	3 g1, 13 a7, 15 h8	0.6°S	24.2°E	7
Monge	15 d4	19.2°S	47.6°E	37
Mons Ampére	5 b6	19°N	4°W	30
Mons Argaeus	1 b5, 3 g6, 4 g1	19°N	29°E	50
Mons Blanc	2 f4	45°N	1°E	25
Mons Bradley	1 g6	22°N	1°E	30
Mons Delisle	5 g8, 6 g2, 8 b3	29°N	36°W	30
Mons Gruithuisen Delta	6 g3, 8 b4	36°N	39°W	20
Mons Gruithuisen Gamma	6 g3, 8 b5	37°N	41°W	20
Mons Hadley	1 f6	27°N	5°E	25
Mons Hansteen	11 e5	12°S	50°W	30
Mons Herodotus	7 d8, 8 e2	27°N	53°W	5
Mons Huygens	5 b6	20°N	3°W	40
Mons la Hire	5 f8, 6 e1	28°N	25°W	25
Mons Maraldi	3 f6, 4 f1	20°N	35°E	15
Mons Pico	2 h4, 6 b4	46°N	9°W	25
Mons Piton	2 g3, 6 a4	41°N	1°W	25
Mons Rümker	8 d5	41°N	58°W	70
Mons Vinogradov	5 g7, 7 a7	22°N	32°W	25
Mons Wolff	5 b5	17°N	7°W	35
Montanari	10 e5, 12 a4	45.8°S	20.6°W	77
Montes Agricola	8 e2	29°N	54°W	160
Montes Alpes	2 f4, 2 g5, 6 a5	46°N	1°W	250
Montes Apenninus	1 g5, 5 a6	20°N	3°W	600
Montes Archimedes	1 h7, 5 b7, 6 b1	26°N	5°W	140
Montes Carpatus	5 f5	15°N	25°W	400
Montes Caucasus	1 f8, 2 e2	39°W	9°E	520
Montes Cordillera	11 h5	20°S	80°W	1,500
Montes Haemus	1 c4, 3 h5	17°N	13°E	400
Montes Harbinger	6 h1, 7 b8, 8 c2	27°N	41°W	90
Montes Jura	6 e5, 8 a6	47°N	37°W	380
Montes Pyrenaeus	15 e5	14°S	41°E	250
Montes Recti	6 d5	48°N	20°W	90
Montes Riphaeus	7 a1, 9 f6, 11 a5	7°S	28°W	150
Montes Rook	11 h2	20°S	83°W	900
Montes Spitzbergen	1 h8, 2 g2, 6 b2, 6 b3	35°N	5°W	60
Montes Taurus	3 e7, 4 e2	25°N	36°E	500
Montes Teneriffe	2 h4, 6 c5	48°N	13°W	110
Moretus	10 c2, 14 f2	70.6°S	5.5°W	114
Morley	15 b7	2.8°S	64.6°E	14
Mortis, Lacus	2 b4, 4 h5	45°N	27°E	150
Mösting	5 b1, 9 a7, 13 h8	0.7°S	5.9°W	25
Mouchez	2 g8, 6 c8	78.3°N	26.6°W	82
Müller	13 f6	7.6°S	2.1°E	22
Murchison	1 g2	5.1°N	0.1°W	58
Mutus	14 c3	63.6°S	30.1°E	78
Naonobu	15 c7	4.6°S	57.8°E	35
Nasireddin	10 b6, 14 f6	41.0°S	0.2°E	52
Nasmyth	12 e4	50.5°S	56.2°W	77
Naumann	8 e4	35.4°N	62.0°W	10
Neander	15 f1, 15 f2, 16 e7	31.3°S	39.9°E	50
Nearch	14 b3, 16 h3	58.5°S	39.1°E	76
Nectaris, Mare	15 f5	15°S	35°E	350
Neison	2 d7	68.3°N	25.1°E	53
Neper	3 a2, 3 a3	8.8°N	84.5°E	137
Neumayer	14 b2, 16 h1	71.1°S	70.7°E	76
Newcomb	3 e7, 3 e8, 4 e3	29.9°N	43.8°E	41
Newton	10 c1, 10 d1	76.7°S	16.9°W	79
Nicolai	14 b6, 16 h5	42.4°S	25.9°E	42
Nicollet	9 c3	21.9°S	12.5°W	15
Nielsen	8 d3, 8 e3	31.8°N	51.8°W	10
Nobili	3 a1, 15 a7	0.2°N	75.9°E	42
Nöggerath	12 d5	48.8°S	45.7°W	31
Nonius	10 a7, 13 f1, 14 f7	34.8°S	3.8°E	70
Norman	9 f6, 11 a5	11.8°S	30.4°W	10
Nubium, Mare	9 d3	20°S	15°W	750
Oceanus Procellarum	7 d3, 7 e6, 8 e1, 8 e3, 11 c8	20°N	55°W	2,000
Oenopides	6 g6, 6 h6, 8 c7	57.0°N	64.1°W	67
Oersted	4 f5	43.1°N	47.2°E	42
Oken	16 c5	43.7°S	75.9°E	72
Olbers	7 h4	7.4°N	75.9°W	75
Olivium, Promontorium *	3 d4	15°N	49°E	2
Opelt	9 d4	16.3°S	17.5°W	49
Oppolzer	1 h1, 13 f7	1.5°S	0.5°W	40

Feature	Map reference	Lat	Long	km
Orontius	10 c6, 14 g6	40.3°S	4.0°W	122
Palisa	9 a5, 13 h6	9.4°S	7.2°W	33
Palitzsch	15 c2, 16 b8	28.0°S	64.5°E	41
Pallas	1 h2, 5 a3, 5 b3	5.5°N	1.6°W	46
Palmieri	11 d2	28.6°S	47.7°W	41
Palus Epidemiarum	9 f2, 10 g7, 12 b7	32°S	27°W	300
Palus Putredinus	1 g7, 2 f1, 5 a7, 6 a1	27°N	0	180
Palus Somnii	3 d4	15°N	44°E	240
Parrot	13 f5	14.5°S	3.3°E	70
Parry	9 c6	7.9°S	15.8°W	48
Pascal	6 e8	74.3°N	70.1°W	106
Peirce	3 c5	18.3°N	53.5°E	19
Peirescius	16 d5	46.5°S	67.6°E	62
Pentland	10 a2, 14 e2	64.6°S	11.5°E	56
Petavius	15 d2, 16 c8	23.3°S	60.4°E	177
Petermann	2 c8	74.2°N	66.3°E	73
Peters	2 d7	68.1°N	29.5°E	15
Petit	3 b1, 3 b2	2.3°N	63.5°E	5
Phillips	15 c2, 16 b8	26.6°S	76.0°E	124
Philolaus	6 c8	72.1°N	32.4°W	71
Phocylides	12 e4	52.9°S	57.3°W	114
Piazzi	11 f1, 12 g7	36.2°S	67.9°W	101
Piazzi Smyth	2 g3, 6 a4	41.9°N	3.2°W	13
Picard	3 c4	14.6°N	54.7°E	23
Piccolomini	13 a1, 14 a8, 15 g2, 16 g8	29.7°S	32.2°E	88
Pickering	13 e7	2.9°S	7.0°E	15
Pico, Mons	2 h4, 6 b4	46°N	9°W	25
Pictet	10 c5, 14 g5	43.6°S	7.4°W	62
Pilâtre	12 e3	60.3°S	86.4°W	69
Pingré	12 e3	58.7°S	73.7°W	89
Pitatus	9 c1, 10 e8	29.8°S	13.5°W	97
Pitiscus	14 b4, 16 h4	50.4°S	30.9°E	82
Piton, Mons	2 g3, 6 a4	41°N	1°W	25
Plana	2 b3, 4 h5	42.2°N	28.2°E	44
Plato	2 g5, 2 h5, 6 b5	51.6°N	9.3°W	101
Playfair	13 e3	23.5°S	8.4°E	48
Plinius	1 b4, 1 c4, 3 g5	15.4°N	23.7°E	43
Plutarch	3 b6, 4 b2	24.1°N	79.0°E	68
Poisson	13 e1, 14 d8	30.4°S	10.6°E	42
Polybius	13 a3, 13 b3, 15 h3	22.4°S	25.6°E	41
Poncelet	6 d8	75.8°N	54.1°W	69
Pons	13 b2, 16 h8	25.3°S	21.5°E	41
Pontanus	13 d2, 14 d8	28.4°S	14.4°E	58
Pontécoulant	16 f3	58.7°S	66.0°E	91
Porter	10 d3	56.1°S	10.1°W	52
Posidonius	1 b7, 2 a1, 2 a2, 3 g8, 4 g3	31.8°N	29.9°E	95
Prinz	7 c8, 8 d2	25.5°N	44.1°W	47
Procellarum, Oceanus	7 d3, 7 e6, 8 e1, 8 e3, 11 c8	20°N	55°W	2,000
Proclus	3 d5	16.1°N	46.8°E	28
Proctor	10 c5, 14 g5	46.4°S	5.1°W	52
Promontorium Agarum	3 h4	14°N	66°E	70
Promontorium Agassiz	2 f3	42°N	2°E	20
Promontorium Archerusia	1 c4, 3 h5	17°N	22°E	10
Promontorium Deville	2 f4	43°N	1°E	20
Promontorium Fresnel	1 f7, 2 e1	29°N	5°E	20
Promontorium Heraclides	6 f4, 8 a5	41°N	33°W	50
Promontorium Kelvin	9 g3, 11 b2	27°S	33°W	50
Promontorium Laplace	6 e5	47°N	26°W	50
Promontorium Lavinium *	3 d4	15°N	49°E	2
Promontorium Olivium *	3 d4	15°N	49°E	2
Promontorium Taenarium	9 b3, 13 h4	19°S	8°W	70
Protagoras	2 e6, 2 f6	56.0°N	7.3°E	22
Ptolemaeus	13 g6	9.2°S	1.8°W	153
Puiseux	9 h3, 11 c2	27.8°S	39.0°W	25
Purbach	9 a2, 13 g2, 14 g8	25.5°S	1.9°W	118
Putredinus, Palus	1 g7, 2 f1, 5 a7, 6 a1	27°N	0	180
Pyrenaeus, Montes	15 e5	14°S	41°E	250
Pythagoras	6 f7, 8 a8	63.5°N	62.8°W	130
Pytheas	5 e6	20.5°N	20.6°W	20
Rabbi Levi	14 b7, 16 h7	34.7°S	23.6°E	81
Raman	7 e8, 8 e2	27.0°N	55.1°W	11
Ramsden	9 g2, 10 h7, 12 c7	32.9°S	3.8°W	25
Rayleigh	3 b7	29.0°N	89.2°E	107
Réaumur	1 g1, 13 f7	2.4°S	0.7°E	53

Feature	Map reference	Lat	Long	km
Recta, Rupes	9 b3, 13 h3	22°S	7°W	110
Recti, Montes	6 d5	48°N	20°W	90
Regiomontanus	9 a1, 10 b8, 13 g2, 14 f8, 14 f9	28.4°S	1.0°W	124
Regnault	8 d7	54.1°N	88.0°W	47
Reichenbach	15 e2, 16 d8	30.3°S	48.0°E	71
Reimarus	16 e4	47.7°S	60.3°E	48
Reiner	7 e4	7.0°N	54.9°W	30
Reiner Gamma *	7 f4	7.5°N	59°W	12
Reinhold	5 f2, 9 d8	3.3°N	22.8°W	43
Repsold	8 d6	51.4°N	78.5°W	107
Rhaeticus	1 f1, 13 e8	0.0°N	4.9°E	46
Rheita	16 e6	37.1°S	47.2°E	70
Rheita, Vallis	16 e6	42°S	51°E	500
Riccioli	7 h2, 11 g8	3.0°S	74.3°W	146
Riccius	14 b7, 16 h6	36.9°S	26.5°E	71
Riemann	4 c5	39.5°N	87.2°E	110
Rima Ariadaeus	1 d2	7°N	13°E	220
Rima Birt	9 b3	21°S	9°W	50
Rima Gay-Lussac	5 e5	13°N	22°W	40
Rima Hesiodus	9 e2	30°S	21°W	300
Rima Hyginus	1 f3	8°N	6°E	220
Rimae Sirsalis	11 f5	14°S	60°W	330
Rimae Triesnecker	1 f2	5°N	5°E	200
Riphaeus, Montes	7 a1, 9 f6, 11 a5	7°S	28°W	150
Ritchey	13 e5	11.1°S	8.5°E	25
Ritter	1 c1, 3 h2, 13 b8	2.0°N	19.2°E	29
Robinson	6 f6, 8 a7	59.0°N	45.9°W	24
Rocca	11 g5	12.7°S	72.8°W	90
Römer	1 a6, 3 f7, 4 f2	25.4°N	36.4°E	40
Rook, Montes	11 h2	20°S	83°W	900
Roris, Sinus	6 g5, 8 c6	51°N	50°W	500
Rosenberger	14 a4, 16 g3	55.4°S	43.1°E	96
Ross	1 c3, 3 h4	11.7°N	21.7°E	25
Rosse	15 f4	17.9°S	35.0°E	12
Rost	10 f3, 12 c3	56.4°S	33.7°W	49
Rothmann	13 b1, 14 a8, 15 h2, 16 g7	30.8°S	27.7°E	42
Rümker, Mons	8 d5	41°N	58°W	70
Rupes Altai	13 b2, 14 a8, 15 h2, 16 g8	24°S	23°E	480
Rupes Liebig	11 d3	25°S	46°W	180
Rupes Recta	9 b3, 13 h3	22°S	7°W	110
Russell	7 f8, 8 g2	26.5°N	75.4°W	103
Rutherfurd	10 d3, 14 g3	60.9°S	12.1°W	48
Sabine	1 c1, 3 h2, 13 b8	1.4°N	20.1°E	30
Sacrobosco	13 c2, 13 c3	23.7°S	16.7°E	98
Santbech	15 e3	20.9°S	44.0°E	64
Santos-Dumont	1 f7, 2 f1	27.7°N	4.8°E	9
Sarabhai	1 c6, 4 h2	24.7°N	21.0°E	8
Sasserides	10 d6, 14 h6	39.1°S	9.3°W	90
Saunder	13 d7	4.2°S	8.8°E	45
Saussure	10 c5, 14 g5	43.4°S	3.8°W	54
Scheele	7 c1, 9 g6, 11 c5	9.4°S	37.8°W	5
Scheiner	10 e3, 12 b2	60.5°S	27.8°W	110
Schiaparelli	7 e7, 8 f1	23.4°N	58.8°W	24
Schickard	12 e5, 12 e6	44.4°S	54.6°W	227
Schiller	10 g4, 12 c4	51.8°S	40.0°W	179
Schlüter	11 h7	5.9°S	83.3°W	89
Schmidt	1 d1, 13 d8	1.0°N	18.8°E	11
Schomberger	14 d1	76.7°S	24.9°E	85
Schröter	5 c2, 9 a8, 13 h8	2.6°N	7.0°W	35
Schröteri, Vallis	7 d8, 8 e2	26°N	51°W	150
Schubert	3 a1, 15 a8	2.8°N	81.0°E	54
Schumacher	4 e5	42.4°N	60.7°E	61
Schwabe	2 c7	65.1°N	45.6°E	25
Scoresby	2 f8	77.7°N	14.1°E	56
Scott	14 d1	81.9°S	45.3°E	108
Secchi	3 d2	2.4°N	43.5°E	23
Seeliger	1 g1, 13 f7	2.2°S	3.0°E	9
Segner	10 g3, 12 d3	58.9°S	48.3°W	67
Seleucus	7 f7, 8 g1	21.0°N	66.6°W	43
Seneca	3 b6, 4 b2	26.6°N	80.2°E	47
Serenitatis, Mare	1 c6, 1 d6, 2 c1, 3 h7, 4 g2, 4 h2	27°N	19°E	650
Shapley	3 c3	9.4°N	56.9°E	23
Sharp	6 g4, 8 b5, 8 b6	45.7°N	40.2°W	40
Sheepshanks	2 d6, 2 e6	59.2°N	16.9°E	25

Feature	Map reference	Lat	Long	km
Short	10 c1, 14 f1	74.6°S	7.3°W	71
Shuckburgh	4 e5	42.6°N	52.8°E	39
Silberschlag	1 e2	6.2°N	12.5°E	13
Simpelius	10 b1, 10 b2, 14 e2	73.0°S	15.2°E	70
Sinas	1 a2, 3 f3	8.8°N	31.6°E	12
Sinus Aestuum	5 c4	12°N	8°W	230
Sinus Iridum	6 e4, 8 a5	45°N	32°W	200
Sinus Lunicus	1 g8, 2 f1, 2 f2	32°N	1°W	100
Sinus Medii	1 g1, 13 f8	3°N	3°E	350
Sinus Roris	6 g5, 8 c6	51°N	50°W	500
Sirsalis	11 f5	12.5°S	60.4°W	42
Sirsalis, Rimae	11 f5	14°S	60°W	330
Smirnov, Dorsa	3 g7	25°N	25°E	130
Smithii, Mare	15 a7	2°S	87°E	360
Smithson	3 c2	2.4°N	53.6°E	6
Snellius	15 d2, 16 c8	29.3°S	55.7°E	83
Somerville	15 b6	8.3°S	64.9°E	15
Sömmering	5 c2, 9 a7, 13 h8	0.1°N	7.5°W	28
Somnii, Palus	3 d4	15°N	44°E	240
Somniorum, Lacus	1 b8, 2 a3, 4 g4	36°N	31°E	230
Sosigenes	1 d3	8.7°N	17.6°E	18
South	6 f6, 8 b7	57.7°N	50.8°W	108
Spallanzani	14 b5, 16 h4	46.3°S	24.7°E	32
Spitzbergen, Montes	1 h8, 2 g2, 6 b2, 6 b3	35°N	5°W	60
Spörer	13 g7	4.3°S	1.8°W	28
Spumens, Mare	3 b1, 15 b8	1°N	65°E	130
Spurr	1 h7, 2 g1, 5 a7, 6 a1	27.9°N	1.2°W	13
Stadius	5 d4	10.5°N	13.7°W	69
Stag's Horn Mountains *	9 b2, 13 h3	26°S	7°W	20
Steinheil	16 f4	48.6°S	46.5°E	67
Stevinus	15 e1, 16 d7	32.5°S	54.2°E	75
Stiborius	14 a7, 15 h1, 16 g7	34.4°S	32.0°E	44
Stöfler	10 a6, 14 e6	41.1°S	6.0°E	126
Stokes	8 d6	52.5°N	88.1°W	51
Strabo	2 b7, 4 h8	61.9°N	54.3°E	55
Straight Wall (Rupes Recta) *	9 b3, 13 h3	22°S	7°W	110
Street	10 d5, 14 h5	46.5°S	10.5°W	58
Struve	7 g7, 8 h1	23.0°N	76.6°W	170
Suess	7 d4	4.4°N	47.6°W	8
Sulpicius Gallus	1 e5	19.6°N	11.6°E	12
Swift	3 c5, 4 c1	19.3°N	53.4°E	11
T. Mayer	5 g5	15.6°N	29.1°W	33
Tacitus	13 c4	16.2°S	19.0°E	40
Tacquet	1 c4, 3 h5	16.6°N	19.2°E	7
Taenarium, Promontorium	9 b3, 13 h4	19°S	8°W	70
Tannerus	14 c3	56.4°S	22.0°E	29
Taruntius	3 d2	5.6°N	46.5°E	56
Taurus, Montes	3 e7, 4 e2	25°N	36°E	500
Taylor	13 c6	5.3°S	16.7°E	42
Tebbutt	3 c3	9.6°N	53.6°E	32
Tempel	1 e2, 13 d8	3.9°N	11.9°E	45
Teneriffe, Montes	2 h4, 6 c5	48°N	13°W	110
Thales	2 b7, 4 h8	61.8°N	50.3°E	32
Theaetetus	1 f8, 2 e2	37.0°N	6.0°E	25
Thebit	9 a2, 13 g3	22.0°S	4.0°W	57
Theon Junior	13 c7	2.3°S	15.8°E	18
Theon Senior	1 d1, 13 c7	0.8°S	15.4°E	18
Theophilus	13 a5, 15 g6	11.4°S	26.4°E	100
Theophrastus	3 e5	17.5°N	39.0°E	9
Timaeus	2 f6, 6 a7	62.8°N	0.5°W	33
Timocharis	5 d7, 6 c1	26.7°N	13.1°W	34
Tisserand	3 d6, 4 d1	21.4°N	48.2°E	37
Tolansky	9 c5, 9 c6	9.5°S	16.0°W	13
Torricelli	3 g1, 15 g7	4.6°S	28.5°E	23
Toscanelli	7 c8, 8 d2	27.9°N	47.5°W	7
Townley	3 b2	3.4°N	63.3°E	19
Tralles	3 d7, 4 d2	28.4°N	52.8°E	43
Tranquillitatis, Mare	1 b2, 3 f3, 13 a8, 15 g8	9°N	31°E	800
Triesnecker	1 g2	4.2°N	3.6°E	26
Triesnecker, Rimae	1 f2	5°N	5°E	200
Trouvelot	2 f5	49.3°N	5.8°E	9
Turner	5 d1, 9 b7	1.4°S	13.2°W	12
Tycho	10 d5, 14 h5	43.3°S	11.2°W	85
Ukert	1 g3	7.8°N	1.4°E	23

Feature	Map reference	Lat	Long	km
Ulugh Beigh	8 g3	32.7°N	81.9°W	54
Undarum, Mare	3 b2	7°N	69°E	220
Väisälä	7 c8, 8 d2	25.9°N	47.8°W	8
Vallis Alpes	2 f5	49°N	3°E	180
Vallis Bouvard	12 h7	39°S	83°W	280
Vallis Rheita	16 e6	42°S	51°E	500
Vallis Schröteri	7 d8, 8 e2	26°N	51°W	150
van Albada	3 b3	9.4°N	64.3°E	22
Van Biesbroeck	8 d2, 8 d3	28.7°N	45.6°W	10
Van Vleck	15 a7	1.9°S	78.3°E	31
Vaporum, Mare	1 f4	13°N	3°E	230
Vasco da Gama	7 h6	13.9°N	83.8°W	96
Vega	16 d5	45.4°S	63.4°E	76
Vendelinus	15 c4	16.3°S	61.8°E	147
Veris, Lacus	11 h5	13°S	87°W	540
Very	1 c6, 3 g7, 4 g2	25.6°N	25.3°E	5
Vieta	11 f2, 12 f8	29.2°S	56.3°W	87
Vinogradov	5 g6, 8 b1	20.0°N	31.1°W	11
Vinogradov, Mons	5 g7, 7 a7	22°N	32°W	25
Vitello	9 g2, 11 c1, 12 d8	30.4°S	37.5°W	42
Vitruvius	1 a4, 3 f5	17.6°N	31.3°E	30
Vlacq	14 a4, 16 g3, 16 g4	53.3°S	38.8°E	89
Vogel	13 e4	15.1°S	5.9°E	27
von Behring	15 b6	7.8°S	71.8°E	39
von Braun	8 f5	41.1°N	78.1°W	62
Voskresenskiy	7 g8, 8 h2	28.0°N	88.1°W	50
W. Bond	2 f7, 6 a7	65.3°N	3.7°E	158
Wallace	5 c6	20.3°N	8.7°W	26
Wallach	1 a1, 1 a2, 3 f2	4.9°N	32.3°E	6
Walter	10 b7, 13 g1, 14 f7	33.0°S	0.7°E	140
Wargentin	12 e5	49.6°S	60.2°W	84
Watt	16 f4	49.5°S	48.6°E	66
Watts	3 d3	8.9°N	46.3°E	15
Webb	3 b1, 15 b7, 15 b8	0.9°S	60.0°E	22
Weierstrass	15 a7	1.3°S	77.2°E	33
Weigel	10 g3, 12 c3	58.2°S	38.8°W	36
Weinek	15 f2, 16 f8	27.5°S	37.0°E	32
Weiss	9 d1, 10 f7, 12 a7	31.8°S	19.5°W	66
Werner	10 a8, 13 f2, 14 f8	28.0°S	3.3°E	70
Whewell	1 e2, 13 c8	4.2°N	13.7°E	14
Wichmann	7 c1, 9 g7, 11 c6	7.5°S	38.1°W	10
Wildt	3 a3	9.0°N	75.8°E	11
Wilhelm	10 e5, 12 a5	43.1°S	20.8°W	107
Wilkins	13 c1, 14 c8	29.4°S	19.6°E	57
Williams	2 a3, 4 g5	42.0°N	37.2°E	36
Wilson	10 f2, 12 b1	69.2°S	42.4°W	70
Winthrop	7 d1, 11 d5	10.7°S	44.4°W	18
Wöhler	14 a7, 16 g6	38.2°S	31.4°E	27
Wolf	9 d3	22.7°S	16.6°W	25
Wolff, Mons	5 b5	17°N	7°W	35
Wollaston	8 d3	30.6°N	46.9°W	10
Wrottesley	15 d3	23.9°S	56.8°E	57
Wurzelbauer	9 d1, 10 e7	33.9°S	15.9°W	88
Xenophanes	8 c7	57.6°N	81.4°W	120
Yakovkin	12 f4	54.5°S	78.8°W	37
Yangel	1 f5	17.0°N	4.7°E	9
Yerkes	3 c4	14.6°N	51.7°E	36
Young	16 e6	41.5°S	50.9°E	72
Zach	10 b3, 14 e3	60.9°S	5.3°E	71
Zagut	13 c1, 14 b8, 16 h7	32.0°S	22.1°E	84
Zähringer	3 e2	5.6°N	40.2°E	11
Zeno	4 d6	45.2°N	72.9°E	65
Zinner	7 e8, 8 f2	26.6°N	58.8°W	4
Zöllner	13 b6	8.0°S	18.9°E	47
Zucchius	10 g2, 12 d3	61.4°S	50.3°W	64
Zupus	11 e4	17.2°S	52.3°W	38